INTRODUCTION TO MARINE MICROPALEONTOLOGY

CONTRIBUTORS

W.A. BERGGREN
ANNE BOERSMA
KRISTER BROOD
LLOYD H. BURCKLE
BILAL U. HAQ
YVONNE HERMAN
LINDA HEUSSER
J. JANSONIUS
W.A.M. JENKINS
STANLEY A. KLING
KLAUS J. MÜLLER
VLADIMÍR POKORNÝ
JÜRGEN REMANE
GRAHAM L. WILLIAMS
JOHN L. WRAY

INTRODUCTION TO MARINE MICROPALEONTOLOGY

EDITED BY

BILAL U. HAQ

ANNE BOERSMA

1998

ELSEVIER

Amsterdam – Lausanne – New York – Oxford – Shannon – Singapore – Tokyo

First edition: 1978

Published by:
Elsevier Science (Singapore) Pte Ltd
No. 1 Temasek Avenue
#17-01 Millenia Tower
Singapore 039192
Singapore

This book is printed on acid-free paper

Printed in The Netherlands

CONTENTS

PHOSPHATIC MICROFOSSILS

ORGANIC-WALLED MICROFOSSILS

PREFACE

There is a resurgence of interest in the subject of Micropaleontology after a period of relative quiescence caused by a slowdown in exploration by the oil industry. Micropaleontology developed as a discipline in the 1940's and 50's when the war effort required more information about the ocean floor and post-war growth in exploration coupled with economic prosperity that resulted in increased science funding. Micropaleontology flourished during the 1960's and 70's when most of the marine microplankton biostratigraphies were developed. This became possible partly due to the availability of the relatively continuous and well-preserved stratigraphic record of the deep sea from all major ocean basins recovered by the Deep Sea Drilling Project and its successor, the Ocean Drilling Program. Higher resolution microscopic and analytical techniques also became available during this period. Accurate age determination and paleoenvironmental/paleoceanographic reconstructions aided by the new biostratigraphic schemes enhanced not only the study of Earth History but also in oil exploration which was perhaps its greatest beneficiary. Exploration was at its peak at that time, and micropaleontologists, armed with their new skills, readily found employment. All of this came to a serious slowdown in the mid 1980s when the price of oil fell below sustainable levels for the industry and new exploration became its first casualty. Industry-wide suspension of new ventures meant loss of interest in allied studies, including biostratigraphic/paleoenvironmental analyses.

In the mid 1990's exploration has begun to pick up once again. More importantly, the development of the new paradigm of Sequence Stratigraphy as a predictive tool for litho- and biofacies has meant a new lease on life for Micropaleontology and biostratigraphy. Sequence Stratigraphy employs the cyclic nature of the stratigraphic record to predict the migration of facies in response to physical and environmental changes. This makes micropaleontological input crucial to producing a realistic chronostratigraphy on which all sequence models depend. It also allows faster and more unequivocal integration of biostratigraphic and paleoenvironmental data in basin analyses and a better understanding of the depositional systems. This application is even more crucial in areas with syndepositional tectonic activity where biostratigraphy becomes a key for correlation and developing regional depositional models. Thus, the renewed interest of the oil industry in the Micropaleontology which should lead to greater enrollments for specialization in the subject at the university level.

Environmental analysis and reconstructions using microfossils benefited both from the more research-supportive economic climate of the 1970's and early 1980's, as well as improved analytical and statistical techniques. Oxygen and carbon stable isotope analyses of several microfossil groups has provided not only environmental information, but also a new chemical stratigraphy of marine deposits and has been a boom for paleoclimatic studies. More recently researchers have been using both organic and inorganic chemical constituents of microorganisms and their shells to interpret their ecology and to extrapolate this information to comprehend world climates. Culturing techniques have also improved since the early, and often unsuccessful, attempts of the 1950's and 60's and now provide key information the habitats of living species. Thus, despite a recent interval of relative slowdown in micropaleontological research, paleoenvironmental analysis is alive and well and in a period of exciting growth.

The rationale for the republication of *Introduction to Marine Micropaleontology* remains the same as that iterated in the preface to the original printing: that there is no comprehensive text on the subject written at the college level. Since the late 1980's, when this volume was allowed to go out of print, several commendable specialized texts have been published, but none of them was able to fill the niche left open by the absence of this text book. They are either too brief or detailed, or their focus is largely on biological or taxonomic aspects of the microplankton groups. *Introduction to Marine Micropaleontology* remains the only succinct text with a truly utilitarian approach, useful not only for an introduction to the subject, but also the application of marine microfossils in paleoceanographic, paleoenvironmental and biostratigraphic analyses. Since the withdrawal of this textbook from book stores, the authors, editors and the publisher have been impressed by the continuing interest of the teachers and practitioners of Micropaleontology in this volume. This, combined with the revival of interest in the subject worldwide led to the decision to reissue the volume in its original, but paperback, form. The ideas expressed in various chapters essentially remain valid today. We have improved the text by updating the Phanerozoic Geologic Time Scale (see page 6 of the introductory chapter) which has been considerably modified since the late 1970s.

We hope that the reprinting of this volume in paperback and its wide availability will help further the cause of Marine Micropaleontology which will continue to be an indispensable member of the family of modern subdisciplines of Earth Science.

March 1997

Bilal U. Haq
Oxford, UK

Anne Boersma
Pomona, NY

ACKNOWLEDGEMENTS

This volume represents the joint effort of many scientists who responded to the editors' invitation to contribute introductory-level chapters in their own fields of specialization to a textbook in micropaleontology. Most contributors received help, advice and permission to reproduce published and unpublished illustrative material from their colleagues. Other colleagues critically reviewed original texts of various chapters, thereby considerably improving their quality. We gratefully acknowledge the cooperation and enthusiastic support of all these colleagues in bringing this book to fruition. Acknowledgements for individual chapters are listed below:

Foraminifera. W.A. Berggren, R. Fleischer, F. McCoy and I. Premoli-Silva for critical reading of the text; J. Aubert, A. Bé, W.A. Berggren, R. Fleischer, H. Luterbacher, D. LeRoy, W. Poag, S. Streeter, and R. Todd for illustrative material and/or permission to reproduce the same; T. Saito and d. Breger for scanning electron micrographs of planktonic foraminifera and A. Edwards and J. Weinrib for help with typing the manuscript.

Calcareous Nannoplankton. D. Bukry, K. Gaarder, K. Perch-Nielsen, R. Poore and H. Thierstein for critical reading of the manuscript. For permission to use illustrative material: A. Farinacci, S. Forchheimer, S. Gartner, S. Honjo, H. Manivit, A. McIntyre and A. Bé; H. Okada, P. Roth, K. Wilbur and N. Watabe also gave permission to use figures.

Ostracodes. H. Oertli for review of the first draft of the manuscript; A. Absolon, D. Andres, R. Benson, H. Blumenstengel, W. van den Bold, K. Debiel, M. Gramm, J. Harding, E. Herrig, N. Hornibrook, V. Jaanusson, A. Keij, R. Kesling, L. Kornicker, H. Kozur, E. Kristan-Tollmann, K. Krömmelbein, H. Malz, A. Martinsson, K. McKenzie, H. Oertli, G. Ruggieri, J. Senes, P. Sylvester-Bradley, Museum of Paleontology, University of Michigan and Senken-Bergische Naturforschende Gesellschaft, Frankfurt, for illustrative material and/or permission to reproduce the same.

Pteropods. G. Tregouboff and M. Rose, Centre National de la Récherche Scientifique, Paris, and the Publisher of Dana Reports for permission to reproduce illustrative material.

Radiolaria. W. Riedel for reviewing the first draft of the manuscript; P. Adshead, H. Forman, B. Holdsworth, H. Ling, E. Merinfeld, T. Moore, Jr., E. Pessagno, W. Riedel, and A. Sanfilippo for illustrative material and/or permission to use the same, and P. Bradley and R. White for scanning electron micro-graphs which were made at the research facilities of Cities Service Oil Company.

Silicoflagellates and Ebridians. D. Bukry and K. Perch-Nielsen for reviewing the manuscript and A. Loeblich, III, Y. Mandra, H. Okada, K. Perch-Nielsen and W. Wornardt for illustrative material and/or permission to reproduce the same.

Dinoflagellates, Acritarchs and Tasmanitids. E. Barghoorn, F. Cramer, E. Denton, J. Deunff, G. Eaton, A. Eisenack, W. Evitt, H. Gorka, T. Lister, A. Loeblich, Jr., F. Martin, W. Sarjeant, D. Wall, D. Williams, Cambridge University Press, Carnegie Institute, Edizioni Tecnoscienza, Harvard Museum of Comparative Zoology, *Micropaleontology*, *Paleontology* and *Revista Española de Micropaleontologia* for illustrative material and/or permission to reproduce the same; J. Charest and M. Trapnell for typing the manuscript and G. Cook for drafting figures.

Spores and Pollen. Y. Tsukada and D. Nicols for illustrative materials.

Chitinozoa. S. Laufeld and Exxon Production Research Co., Houston, for illustrative material.

1

MARINE MICROPALEONTOLOGY AN INTRODUCTION

W.A. BERGGREN

By definition **micropaleontology**, the study of microscopic fossils, cuts across many classificatory lines. It includes within its domain the study of large numbers of taxonomically unrelated groups united solely by the fact that they must be examined with a microscope. At the same time within certain taxonomically homogeneous groups the size of some forms is such that they scarcely need be examined with microscopic aid and are more properly grouped under macropaleontology. It is not surprising then that as a discipline micropaleontology lacks a certain coherent homogeneity. Most marine microfossils are protists (unicellular plants and animals), but others are multicellular or microscopic parts of macroscopic forms. Thus, their grouping into one discipline remains essentially practical and utilitarian.

The practical value of marine microfossils in various fields of historical geology is enhanced by their minute size, abundant occurrence and wide geographic distribution in sediments of all ages and in almost all marine environments. Due to their small size and large numerical abundance, relatively small sediment samples can usually yield enough data for the application of more rigorous quantitative methods of analysis. Moreover, most planktonic and many benthic microfossils have wide geographic distributions that make them indispensable for regional correlations and comparisons, and paleooceanographic reconstructions.

Marine microfossils occur in sediments of Precambrian to Recent ages, and in every part of the stratigraphic column one or more groups can always be found useful for biostratigraphic and paleoecologic interpretations (see Fig. 1). Terminology for the major divisions of the marine realm discussed in this text are shown in Fig. 2. Fossil marine organisms lived in almost all these marine areas and can therefore be invaluable in the study of changes in the paleoenvironments.

For instance, radiolaria, silicoflagellates, calcareous nannoplankton, pteropods, and some foraminifera and diatoms (Fig. 1) are planktonic (i.e. free floating) and live in abundance from 0 to 200 m in the open ocean, but diminish rapidly near the continents. These forms are useful in monitoring past changes in the oceanic environments, particularly changes in temperature. Other groups (Fig. 1) such as the ostracodes, bryozoa, and some foraminifera and diatoms are benthic (i.e., adapted to living on the bottom of the sea), as either **vagile** (free-moving) or **sessile** (passive or attached) organisms. Since these forms exhibit distribution patterns broadly linked to depth, sediment type and various physico-chemical variables in seawater, they are useful in delineating changes in the bottom environment.

Some forms, such as the dinoflagellates, are known to contain both planktonic and benthic phases in their reproductive cycle and are particularly useful tools in paleo-

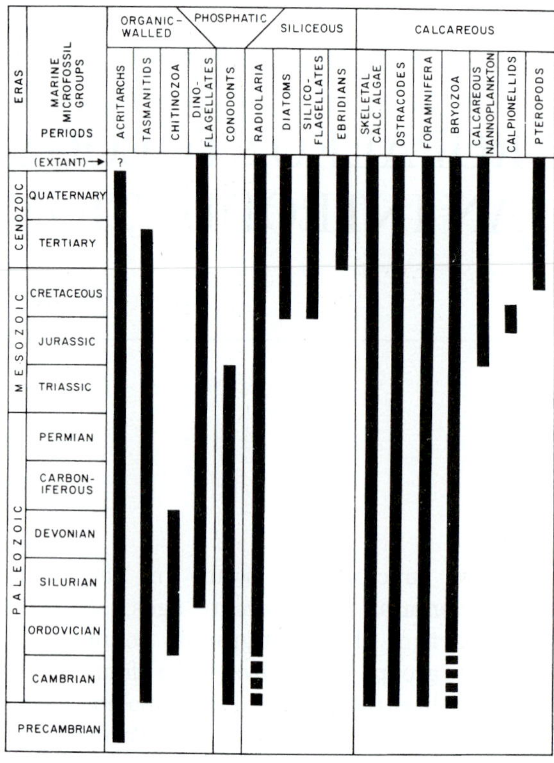

Fig. 1. Stratigraphic distribution of the major marine microfossil groups.

ecologic studies of near-shore areas and inland seas (e.g., the Red Sea, Black Sea, etc.). Others, such as the conodonts (extinct since the early Mesozoic) are presumed to have been attached to the soft body of a planktonic animal and thus their use in reconstructing Paleozoic and early Mesozoic planktonic ecology is more tenuous than the use of forms with living descendants in the Mesozoic and Cenozoic.

Spores and pollen, although derived from land plants, are strongly climate-dependent. Thus, their presence and distribution patterns in near-shore marine sediments allow interpretations of continental climates; and/or, like chemical tracers, their distribution can be used to monitor current movements.

HISTORICAL REVIEW

The earliest mention of microfossils dates back to classical times. The large benthic foraminifer *Nummulites* — the great rock builders of the pyramids — had been mentioned by Herodotus (5th century B.C.), Strabo (7th century B.C.) and Pliny the Elder (1st century A.D.). While in the western world we usually credit Leonardo da Vinci (1452–1519) as having been the first to recognize the organic nature of fossils, it would appear that Leonardo had been anticipated by some three centuries by one Chu Hsi, who in his *Analects*, dated 1227, actually perceived their true nature.

The systematic study of microfossils awaited discovery in 1660 by Antonie van Leeuwenhoek of the microscope and it would seem that this is an appropriate date to denote the birth of systematic micropaleontology. The foraminifera were the first group of microfossils to receive the attention of early naturalists. Although illustrated in various publications as early as the 16th century and given short Latin descriptive diagnoses, it was not until the 10th edition of Linné's *Systema Naturae* (1758) that a binomial nomenclature, which was to form the basis of modern biological taxonomy, was applied to some fifteen species of foraminifera.

Alcide d'Orbigny's (1802–1875) detailed studies on the foraminifera led to the first comprehensive classification of the foraminifera in 1826 (he included them among the cephalopods) and subsequently published a large number of papers on living and fossil foraminifera. He was also first to utilize foraminifera in biostratigraphic studies of the Paratethyan Tertiary. In 1835 the French biologist Felix Dujardin demonstrated that the foraminifera could not be considered as cephalopods because they were characterized by long pseudopodia and he introduced the name Rhizopodes for the group. D'Orbigny and most of his contemporary colleagues accepted this revision. Not however C.G. Ehrenberg (1795–1876), a German micropaleontologist and contemporary of d'Orbigny. Ehrenberg was a remarkable scientist of the period and is generally credited with having made the first discovery and description of silicoflagellates, ebridians, coccoliths, discoasters, dinoflagellates, and numerous living protists, in addition to having described and illustrated numerous radio-

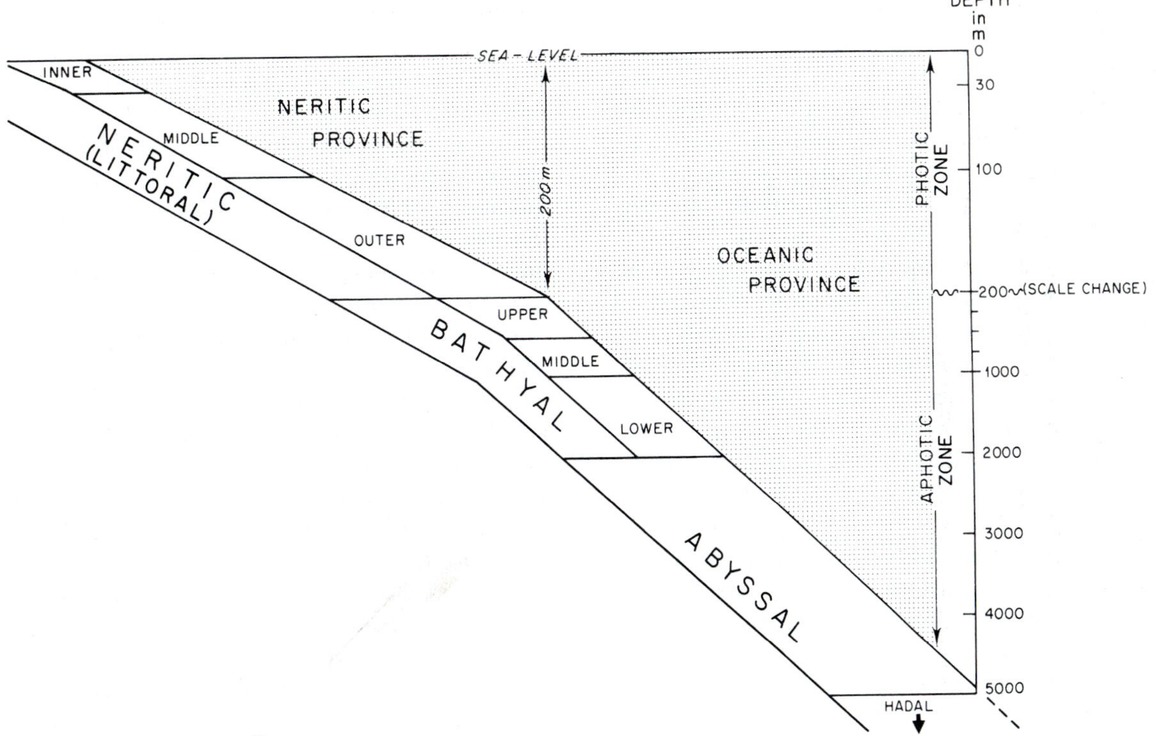

Fig. 2. The main divisions of the marine environment.

larians, diatoms and foraminiferans. Ehrenberg rejected Dujardin's views and claimed that the foraminifera belonged to the Bryozoa, a view which he staunchly maintained as late as 1858.

Basic descriptive studies were conducted on many of the major microfossil groups during the second half of the 19th century. The studies of Reuss in the 1860's and 1870's on the Cretaceous and Tertiary foraminifera of Prussia, though contemporaneous, contrast strongly with those of the so-called "English school" which included N.C. Williamson, W.K. Parker, T.R. Jones, W.B. Carpenter, H.B. Brady, and C.D. Sherborne. The latter group were distinguished by their view that foraminifera exhibited such large individual variation as to make specific differentiation virtually impossible, thus rendering the group of little use in stratigraphic studies, a view that, unfortunately, remained influential well into the present century.

At the same time the major studies by Ernst Haeckel on the Radiolaria (1862–1887), the basic taxonomic work on ostracodes by Sars (1866), and the Schmidt diatom atlas (beginning in 1875), among others, testify to the vigorous descriptive work conducted in micropaleontology during this time. After this auspicious beginning, however, it is curious that interests subsequently waned in all these groups save the foraminifera.

The single largest impetus to descriptive micropaleontologic studies in the latter half of the 19th century was the voyage of H.M.S. *Challenger* from 1873 to 1876. In 1884 H.B. Brady published a monumental monograph of the foraminifera dredged by the *Challenger* during its voyage around the world. This work remains to this day the fundamental reference in the study of Neogene and living foraminiferal faunas.

The second, or analytic, phase in micropaleontology is concerned with such things as basic biologic structure and affinities, morphogenetic studies, creation of classificatory systems approaching a "natural system" and biostratigraphic applications.

The single major cause for a revival of interest in marine micropaleontology and which has led directly to the third or synoptic phase in most groups is the advent of oceanographic deep sea drilling. Oceanographic micropaleontology expanded greatly in the post-World War II period with the development of better techniques for obtaining deep-sea piston cores, particularly in the 1950's and 1960's. This was substantially aided by the advent of the JOIDES Program in 1965 and its successor, the Deep Sea Drilling Project (DSDP), in 1968. With the conclusion of this third phase of ocean drilling in autumn 1975, the D/V *Glomar Challenger* had drilled over 400 sites in all areas of the world oceans except the Arctic. It recovered over 41,000 m of nearly 76,000 m of drilled and cored sediments and basalt dating back to the mid-Jurassic in water depths from less than 1 to more than 4 km. The availability for study of sediment from the ocean floors covering the last 150 million years has allowed great advances in the study of the evolution of the oceanic planktonic and benthic microfossils and the utilization of these forms in deciphering the history of the evolution of the oceans themselves during this time. Refined biostratigraphic zonations and paleobathymetric and paleoecologic studies are but a few areas in which great strides have been made in the past decade. These synoptic studies were not possible until a sufficient data base was developed — a data base which had to await a sufficient amount of analytical study on the taxonomic groups themselves and the availability of a sufficient amount of material from wide areas of the marine biosphere. The oceanic region until recently remained the last great frontier for the micropaleontologist.

Commercial micropaleontology

Micropaleontology as a relatively uniform discipline is a young field of science dating back to the First World War. At that time the need for greater amounts of petroleum led to the enhanced study of foraminiferal faunas as a correlation tool. The conversion from what had previously been the somewhat esoteric descriptive studies on foraminiferal taxa to the utilitarian aspects of biostratigraphic correlation in search for greater amounts of fossil fuels was rather rapid.

Commercial, or industrial, micropaleontology is generally associated with the petroleum industry. With the recognition of the biostratigraphic utility of foraminifera in petroleum exploration, micropaleontology received a new impetus and direction. Micropaleontology continues to be widely used in the exploration and mapping of surface and subsurface stratigraphic units by geological surveys all over the world. At the same time the expansion of commercial micropaleontology has been intimately linked with its development as a part of the academic curriculum in the earth sciences. In 1911 Professor J.A. Udden of Augustana College (Rock Island, Ill.) stressed the importance of using microfossils in age determination and correlation in subsurface water well studies in Illinois. At the same time he convinced oil companies of the importance of microfossils in age determination of drill cuttings. During World War I, micropaleontology was introduced as a formal course in the college curriculum by Josia Bridge at the Missouri School of Mines and by H.N. Coryell at Columbia University and by F.L. Whitney at the University of Texas (Austin). The latter taught Hedwig Kniker, Alva Ellisor, and Esther Applin. These three women were subsequently to become the leading economic micropaleontologists in the United States and Ellisor eventually became the chief paleontologist for Humble Oil and Refining Company.

J.J. Galloway and J.A. Cushman were prime movers of micropaleontology at this time. The year 1924 was an important one in micropaleontology. Galloway began teaching micropaleontology at Columbia University. In the same year the Cushman Laboratory for Foraminiferal Research was established at Sharon, Massachusetts, and was affiliated with Harvard University. For the next quarter of a century this laboratory remained one of the major centers of foraminiferal research in North America and many of the micropaleontologists who went on to successful careers in industry and universities were at one time or another associated with J.A. Cushman. Micropaleontology was introduced at Leland

Stanford University in 1924 by H.G. Schenck and in 1933 the study of ostracodes was introduced at Columbia University by H.N. Coryell. C.G. Croneis introduced micropaleontology at the University of Chicago during the depression. In the decade immediately preceding World War II emphasis in micropaleontology and petroleum exploration remained heavily concentrated in biostratigraphic correlation, whereas in the decade immediately following the war emphasis shifted rapidly to paleoecology and paleobathymetry.

BASIC CONCEPTS AND REVIEW OF CURRENT TRENDS

Biostratigraphy and biochronology

Interpretations of earth history depend on two different systems of logic, both of which arrange geological observations into sequences of events. The first and most widely used is the logic of superposition: the ordering of events *iteratively* in a system of invariant properties simply by determining the physical relationship of features in the rocks. This is what is meant by the word **stratigraphy**. The second logical system depends on the recognition of an *ordinal* progression which links a series of events in a system of irreversibly varying characteristics. This provides a theoretical basis outside of the preserved geological record by which the nature and relationship of the events in the progression can be recognized or predicted, and according to which missing parts of the record can be identified. Geology is an historical philosophy, so the ordinal progressions we refer to are progressions in time, just as geological time is perceived by the progress in one or another ordinal series of events. This is what is meant by the word **geochronology**. A time-scale then provides a conceptual framework within which to interpret earth history; in this specific instance, phenomena related to the evolution of the marine biosphere. Only two ordinal scales are widely used today, that of **radiochronology** (based on isotope-decay rates) and that of **biochronology** (based on organic evolution). Geomagnetic polarity reversals are non-ordinal repetitions but because of their wide applicability have been closely calibrated to the ordinal time-scale. The geologic column and its radiochronologic framework used as the basis for discussions in this text is shown in Fig. 3.

Philosophy and methodology behind the establishment of a time-scale

Stratigraphy has been succinctly defined as the "descriptive science of strata". It involves the form, structure, composition, areal distribution, succession and classification of rock strata in normal sequence. **Biostratigraphy** is that aspect of stratigraphy which involves the direct observation of paleontologic events in superposition. A biostratigraphic unit is a body of rocks which is delimited from adjacent rocks by unifying contemporaneous paleontologic characteristics.

The evolution of organisms through time has provided the framework for a system of zonations by which discrete units of time represented by material accumulation of sediments can be recognized. Biozones may generally be grouped into three categories depending on their characteristic features: (1) **assemblage zones**, those in which strata are grouped together because they are characterized by a *distinctive natural assemblage* of an entirety of forms (or forms of a certain kind) which are present; (2) **range zones**, those in which strata are grouped together because they represent the *stratigraphic range* of some *selected element* of the total assemblage of fossil forms present; and (3) **acme zones**, those in which strata are grouped together because of the *quantitative presence (abundance)* of certain forms, regardless of association or range. The latter are, qualitatively, of lesser importance than the first two. Most planktonic zones used in current biostratigraphic work are range zones and are of the following types:

(a) **Taxon Range Zone** — a body of strata representing the total range of occurrence (horizontal and vertical) of specimens of a taxon (species, genus, etc.).

(b) **Concurrent Range Zone** — a range zone

FIGURE 3 THE PHANEROZOIC GEOLOGIC TIME SCALE

ERA	PERIOD		EPOCH	AGE	BOUNDARY AGE (in Ma)
Cenozoic	Quaternary		Pleistocene-Holocene		0
	Tertiary		Pliocene		1.8
			Miocene		5.3
			Oligocene		24
			Eocene		34
			Paleocene		55
MESOZOIC	Cretaceous		Late	Maastrichtian	65
				Campanian	
				Santonian	
				Coniacian	
				Turonian	
				Cenomanian	
			Early	Albian	
				Aptian	
				Barremian	
				Hauterivian	
				Valanginian	
				Berriasian	
	Jurassic		Late	Tithonian	145
				Kimmeridgian	
				Oxfordian	
				Callovian	
			Mid	Bathonian	
				Bajocian	
				Aalenian	
			Early	Toarcian	
				Pliensbachian	
				Sinemurian	
				Hettangian	
	Triassic		L	Rhaetian	206
				Norian	
			M	Carnian	
				Ladinian	
			E	Anisian	
				Scythian	
PALEOZOIC	Permian		L	Ochoan	248
				Guadalupian	
				Leonardian	
			E	Sakmarian	
				Asselian	
	Carboniferous	Pennsylvanian	L	Stephanian	295
				Westphalian	
				Namurian	
		Mississippian	E	Visean	
				Tournaisian	
	Devonian		L	Famennian	354
				Frasnian	
			M	Givetian	
				Eifelian	
				Emsian	
			E	Pragian	
				Lochkovian	
	Silurian		L	Pridolian	417
				Ludlovian	
			E	Wenlockian	
				Llandoverian	
	Ordovician		L	Ashgillian	443
				Caradocian	
			M	Llandeilian	
				Llanvirnian	
			E	Arenigian	
				Tremadocian	
	Cambrian		L	Shidertinian	495
				Tuorian	
				Mayan	
			M	Amgan	
				Lenan	
			E	Atdabanian	
				Tommotian	
				Manykaian	
					545

Scale: 0, 50, 100, 150, 200, 250, 300, 350, 400, 450, 500 — millions of years (B.P.)

(Updated 1997)

defined by those parts of the ranges of two or more taxa which are concurrent or coincident. It is based on a careful *selection* (and rejection) of faunal elements which have a concurrent, though not necessarily identical, stratigraphic range with a view to achieving a biostratigraphic unit of maximum time-discrimination and extensibility.

(c) **Oppel Zone** — a zone characterized by a distinctive association or aggregation of taxa selected because of their restrictive and largely concurrent range, with the zone being defined by the interval of common occurrences of all or a specified portion of the taxa. This is a less precise, and more restricted, relative of the Concurrent Range Zone described above. It is little used in planktonic biostratigraphy. As a biochronological concept, it is exemplified by the Land Mammal Age.

(d) **Lineage Zone** or **Phylozone** — the body of strata containing specimens representing the evolutionary or developmental line or phylogenetic trend of a taxon or biologic group defined above and below by features of the line or trend. It has been commonly referred to as a **phylogenetic zone**.

The scope of a lineage zone (phylozone) may extend from the first (evolutionary) appearance of some form in an evolutionary bioseries to the termination of the lineage, thus including the whole bioseries or lineage, or it may include only a segment of the lineage **(lineage-segment zone)**. For a further discussion on the nature and application of the lineage-zone concept to biostratigraphic studies the reader is referred to Van Hinte (1969) and Berggren (1971).

(e) **Acme Zone** — a body of strata representing the acme or maximum development of some species, genus or other taxon, but not its total range. It is little used, except in local biostratigraphy in planktonic studies.

(f) **Interval Zone** — the interval between two distinctive biostratigraphic horizons but not in itself representing any distinctive biostratigraphic range, assemblage or feature. The so-called Partial-Range Zone actually corresponds to an interval zone and is commonly used in current planktonic biostratigraphic studies. For a more comprehensive discussion on the nature of biostratigraphic zones the reader is referred to the *Preliminary Report on Biostratigraphic Units* published by the International Subcommission on Stratigraphic Classification, Report No. 5 at the 24th International Geological Congress, Montreal, 1972.

Biostratigraphic zones differ in a fundamental manner from time-stratigraphic units (see Fig. 4). The latter includes, by definition, all the rocks formed during a given interval of geologic time. Biostratigraphic zones as defined are limited to the areal distribution of strata which actually contain specific fossil forms or assemblages. No biostratigraphic zone is of mondial, absolutely synchronous extent, because no fossil form had an instantaneous mondial origin nor suffered instantaneous extinction, nor was so independent of environment as to be found everywhere in all types of sediments laid down synchronously.

In the area of marine biostratigraphy perhaps the best examples of biostratigraphic zones are those proposed for Mesozoic and Cenozoic sequences based upon calcareous and siliceous plankton. From initial studies on the early Cenozoic planktonic foraminifera of the SW Soviet Union in the decade preceding World War II, through similar studies on the Cenozoic of the Caribbean region following the war, to the advent of the Deep Sea Drilling Project in the mid-late 1960's we have witnessed the creation of a veritable flood of biostratigraphic zonations within the Cenozoic at least, and particularly for its younger part — the Neogene — local and regional biostratigraphic zonations have been created for virtually all areas of the marine realm from equator to subpolar regions reflecting the dependency of such zonations upon biogeography (i.e. climatology — paleooceanography).

Obviously, the radiochronological time-scale can be quantified since it is based on well-documented assumptions of the invariancy of isotope-decay rates through geological time; and since radiometric ages are expressed in numbers, they conform to what we are conditioned to accept as measurement itself.

The methodology and reasoning by which the marine time-scale has been developed is as follows:

(1) From 0 to 5 Ma (million years before present) the time-scale is based upon the radiometrically controlled paleomagnetic time-scale of Cox (1969) and the calibration

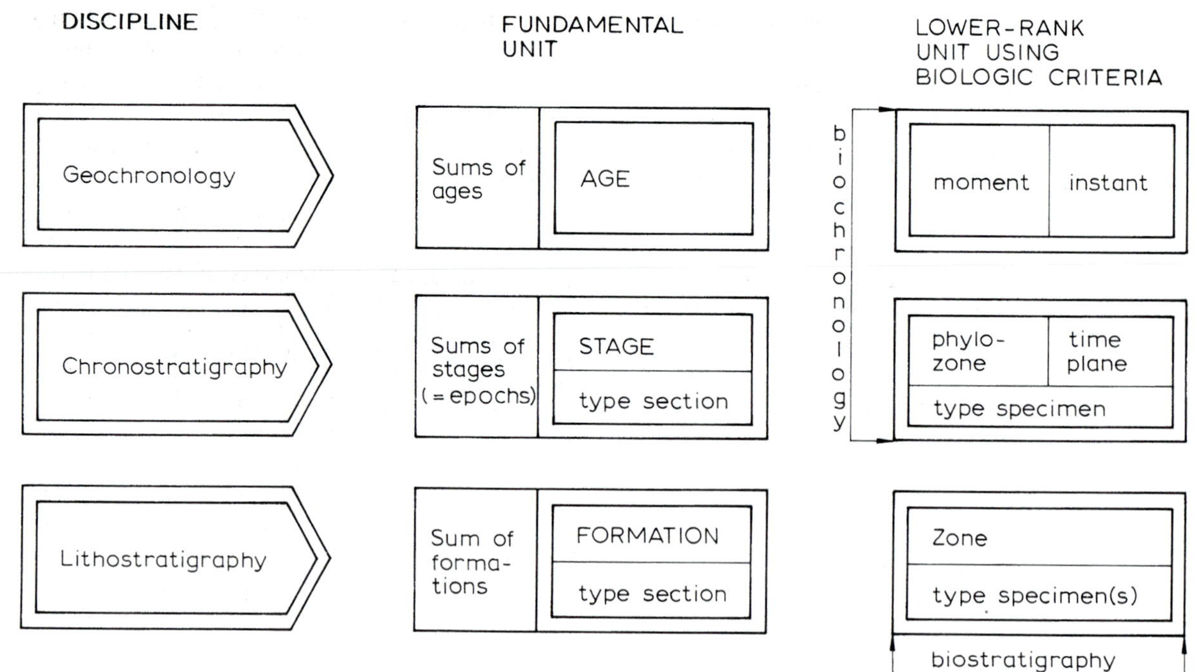

Fig. 4. Relationship between the discipline of geochronology, chronostratigraphy, and lithostratigraphy. Geochronological events are placed directly above their chronostratigraphic equivalents.

Geochronology, the measuring of time, provides an absolute, ordinal scale, or framework, for the geological column. Geochronology further aims at depicting the time of occurrence of a given event such as an event involving a fossil organism or a magnetic anomaly. The fundamental unit for subdividing the geochronological scale is the *age*. In turn, ages may be subdivided based on some biologic criterion which is relatable to absolute time. Thus, a *moment* is the time during which a taxon is known to have existed (but due to our imperfect knowledge of many organisms, may not be its true total time of existence).

Chronostratigraphy, on the other hand, relates sequences of rock units to time. The major unit in chronostratigraphy is the *stage* which describes strata lain down during a specified period of time. A stage is based on a designated locality, its **type section**, to which all equivalent rocks can then be related. For example, the type section of the Maastrichtian Stage is near Maastricht, The Netherlands; rocks considered to be that age in England can then be compared, and if equivalent, are called Maastrichtian in age. A stage may further be characterized by biological events concerning its incorporated fauna which are related to time. Events such as the total time existence of a taxon (species), called its **phylozone**, or a single event occuring at a specific time and measurable over some distance, the **time** plane, fall into the realm of chronostratigraphic biochronology. Due to the rarity of world-wide synchroneity of events in the biosphere, just now being appreciated by biostratigraphers, time planes are difficult to prove on the basis of fossils.

Lithostratigraphy, by distinction, deals with sequences of rocks, *independently of time*. The primary unit for categorizing rock units is the **formation**, which also is based on a type section. Lithostratigraphic units describe only strata, not the time involved in their deposition. The biostratigraphic **zone** is then a biological characterization of part or all of arock unit; such divisions may be based on one of several criteria such as the first occurrence of a taxon in a section, overlapping ranges of two taxa, etc. This type of zone does not specify the time of an event, as the same event may be diachronous (non-time equivalent) between two geographic regions, in two parts of an ancient basin, or in environments which change at different rates.

Along these lines, consider the problems involved in relating low-latitude to high-latitude zonations. (Modified after Van Hinte, 1969).

to it of biostratigraphic events which have been achieved by investigations on deep-sea cores. The calibration of this part of the time-scale is believed to have a high degree of accuracy and reliability.

(2) From 5 to 25 Ma the time-scale is based upon calibration of biostratigraphic events to K–Ar dates, the calibration to the paleomag-

netic time-scale of selected biostratigraphic events in certain fossil plankton groups, particularly the radiolaria, and thence by second-order correlation between various planktonic microfossil groups, and to correlations between the sea-floor magnetic anomalies and paleomagnetic epochs. The calibrations within this part of the time-scale must be viewed as moderately tentative.

(3) Beyond 25 Ma the time-scale is based upon calibration of biostratigraphic events and the geomagnetic reversal record to K–Ar dates, interzonal correlations and linear extrapolations between biostratigraphic events and/or zonal boundaries in cores and outcrops. The calibrations within this part of the time-scale are considered to be tentative and subject to moderate revision as the paleomagnetic time-scale is gradually extended through the Paleogene. Current studies on the late Cretaceous and early Cenozoic (Paleocene) paleomagnetic stratigraphy of land sections and correlation with the sea-floor magnetic anomalies, as well as calibration of biostratigraphic events to K–Ar dates at the basaltic basement/sediment contact, appear to offer calibration control points upon which the time-scale may be extended over the lower part of the Cenozoic (Paleogene) and into the Mesozoic.

The major problem in creating a relatively accurate time-scale for the Mesozoic is the difficulty in matching the paleomagnetic reversal sequence in land sections with the magnetic anomaly pattern on the sea floor and their calibration to K–Ar dates, owing to the large experimental error in radiometric dates of this antiquity. Considerable progress has been made in the last few years, so that despite current controversies associated with the developing Mesozoic time-scale, we can look with optimism to a relatively reliable Mesozoic time-scale within the next few years.

The pre-Mesozoic time-scale is based solely on the correlation between biostratigraphic events and K–Ar dates with ever increasing uncertainty limits with increasing antiquity.

There is a persistent movement among earth scientists, understandably among specialists in physical geology and mineralogy but also among paleontologists who work with the more "datable" parts of the fossil record, to do away with biochronology as a primary means of dividing up geological time and to rely entirely on arbitrary and unquestionably convenient time-lines, expressed in rounded-off and exact numbers of years, which are to be identified in the geological record by radiochronology. Biochronological age measurements, even though they are not as easily quantified, are nevertheless implicit in the fossil record itself and do not systematically lose resolution in the older parts of that record as radiometric ages do. Arguments as to the real meaning of radiometric numbers, and as to their practical utility everywhere, will eventually divide on the degree to which (numerical) certainty can be demanded in an uncertain universe, but it is our opinion that "biological time" — that is the history of the biosphere and its physical environment — is best measured biochronologically with a critically evaluated calibration assist wherever possible from radiochronology. To do otherwise is to let the radiometric tail wag the paleontological dog.

Biology and evolution

Evolutionary theory provides the conceptual framework for biological systems just as time provides the conceptual framework for (bio)stratigraphy. Patterns of fossil "behavior" such as diversity, geographic distribution, community structure, intraspecific morphologic variability, can be understood in the context of evolutionary theory. However, the path between micropaleontologic studies and evolutionary concepts is a two-way street: whereas evolutionary theory may synthesize and render more concrete (i.e. objective) the patterns of fossil "behavior", the data from micropaleontologic studies on living or fossil populations provide the raw material upon which evolutionary theory is based and continuously subjected to modification.

Evolutionary concepts applied to (micro) fossil populations are based upon studies of the biology and ecology of living organism populations which points out the strong tie between evolution and ecologic studies. Evolutionary theories are based, in turn, upon the science of genetics, the precepts of which are rarely studied by micropaleontologists. Nevertheless, a basic understanding of evolution and genetics is essential to the fields of (paleo)ecology and biostratigraphy. For instance, Simpson (1953, p. 138) has defined selection as "anything tending to produce systematic, heritable change in populations between one generation and the next". This broad, modern definition may be contrasted with the limited, restricted mid-nineteenth century Darwinian concept of natural selection as the selective reproduction by the most successful members of a population leading to adaptive changes in species and ultimately to new species. Selection is the mechanism of interaction between organism and environment and, as Simpson has pointed out, its role in evolution is the production of adaptation. The marine microorganisms discussed in this book represent some of the most successful adaptations in the history of life in terms of sheer numbers, diversity, geologic record (longevity), ecologic differentiation and evolutionary radiation.

The concept of the "species" is of critical importance to (micro)paleontology, although in fossil populations the conditions of the biologic definition cannot be met. Mayr (1966) defines a species as "groups of interbreeding natural populations that are reproductively isolated from other such groups".

In contrast the paleontologic species concept is more subjective and utilitarian: a species is defined primarily on the basis of a set of external, preservable morphologic features which can be recognized and applied by other specialists. Subjectively interpreted antecedent and descendant relationships and stratigraphic range form an integral part of the definition and recognition of a fossil taxon.

Whether a group of organisms reproduces sexually or, as most of the forms discussed in this text, asexually, further affects the definition (i.e. characterization) of a fossil species and the taxonomy used to describe it. Some asexually reproducing organisms are capable of extensive morphologic variability which will affect their classifications and nomenclature. Still others may assume distinctly different morphologies during the course of ontogeny (e.g. dinoflagellates vs. hystrichomorphs) and hence present their own specific problems of classification and nomenclature. These and other patterns of fossil "behavior", whether it be within a given lineage, intraspecific variation due to environmental variables, large-scale adaptive radiations, or individual ontogenetic development, affect the interpretations made by the micropaleontologist.

It should be apparent that a given taxon cannot be adequately denoted by reference to a type specimen (or specimens) alone. A number of specimens should be utilized to define the concept of a species (or other lower taxonomic category) and the term **hypodigm** has been suggested by Simpson (1940) for those individuals chosen collectively from the original material which are considered to be representative of the population of the new taxon. Yet, all too often we read in past and current literature of slight differences which a specimen (or a few specimens) has from the holotype, the type specimen, of another species and of the consequent "necessity" of erecting a new species on this basis. Such static, rigid concepts cannot be applied to highly variable members of an evolutionary continuum. Only by recourse to a number of individuals (the more the better) can the intrapopulation variation of a given taxon be determined with satisfactory accuracy. The decision as to whether a given species is conspecific with another is, in the last analysis, based on the inferred range of variation for the whole taxon as a result of studying adequate comparative material, not on the morphologic similarity to any given specimen (holotype, or other). It is the population and inferences drawn from it, not the holotype specimen itself, which must serve as the ultimate criterion of identification and, ultimately, classification of a taxon. The typological approach is inextricably linked with philosophical idealism, as Simpson (1961) has pointed out, and as such has no place in modern paleontologic thought.

It is generally accepted today that classification should be based upon phylogeny. We may paraphrase Simpson (1961) in listing several criteria of strictly Darwinian taxonomy which should be utilized in attempts to erect a classification founded on phylogeny:

(1) Taxonomic groups are the results of des-

cent with modification, or phylogeny.

(2) Each valid taxon has a common ancestry.

(3) The fundamental, but not sole, criterion for ranking of taxa is propinquity of descent.

Populations, not the characters observable on the individual forms, are the things classified. The characters chosen to define the taxa are to be interpreted as showing evidence of phylogenetic affinities and to be ranked in accordance with their probable bearing on nearness of descent. The same character may in one group characterize a genus or a family, whereas in another a species or subspecies These characters are not *a priori* determinable, but their rank must be ascertained as a result of experience. The individual is merely referred, by inference, to a population of which it is a small (in the case of most marine microorganisms an infinitesimal) part. The variation which one observes in populations of different species is an inherent part of their nature and definition. Types serve merely as the legislative requirement of nomenclature. Observations of individual morphology and other somatic attributes of the species will aid in determining whether the evolutionary definition is met by a given population.

Even in instances where a strict application of quantitative data is not made, the fundamental approach of taxonomic studies is statistical in nature. This is because the observations on populations in nature can, at best, provide only partial information on the taxon of which it is a member. It is not the characters and general similarity between individuals which are of primary significance in determining membership in a given taxonomic category, but rather the *relationships* which these characters express which are of main importance. These relationships are evolutionary; in short, they are phylogenetic.

Supraspecific taxa are delimited on the principal of monophyly — all members of a taxon having a single phylogenetic origin. The pragmatic criterion of monophyly is derivation from an ancestral taxon of the same, or lower, rank (see Simpson, 1961).

Consider also the difference in material available to the specialist working on Recent faunas and the micropaleontologist working with fossil forms. Where the former has access through accurately collected plankton tow material and ocean bottom sediment cores to a suite of specimens representing all shades of morphologic development from genesis through gerontism, the paleontologist is limited to death assemblages — **thanatocoenses**. Thus, where the Recent plankton specialist has an advantage in making observations on living populations and drawing the most logical inferences with regard to the taxonomic composition of Recent faunas, he is at the same time at a distinct disadvantage in utilizing these observations as the foundation of a coherent, logically founded classification. The reason for this is the time element which is lacking in his studies. For example, the calcareous and siliceous planktonic species in the present-day oceans are the result of a long sequence of phylogenetic events in the Tertiary, and, as classification is to be based upon phylogeny, only by taking cognizance of these events can a satisfactory classification of these organisms be formulated.

Phyto- and zoogeographic data are of further importance in providing information on taxonomy and classification. Ecologic and paleoecologic data may provide criteria whereby the difficult problem of distinguishing between convergence (which implies similar ecologic conditions among forms of unrelated phylogenies) and parallelism (which refers to the independent acquisition of similar structure in forms having a common genetic origin) may be resolved.

In the elucidation of characters distinctive of a given specific or subspecific taxon, geographic variation in the character — the **geographic cline** concept — may provide a clue as to the nature and limits of variation within that taxon. It is primarily an evolutionary and not a taxonomic concept and refers to an intraspecific gradation in measurable characters. Various types of clines have been recognized, such as the geocline (geographic), ecocline (ecologic), chronocline (successional). Examples of geoclines in the Cenozoic planktonic foraminifera include the latitudinal modification of the umbilicus and umbilical collar and the height of the conical angle (i.e. degree of convexity) within a given species and between successional species (chronocline); variation in development of intra-umbilical "teeth" in the species *Neogloboquadrina dutertrei*. The most obvious example of an Upper Cretaceous geographic cline is the marked difference in surface ornament in the planktonic foraminifer *Rugoglobigerina* from spinose to hispid in high latitudes to meridionally oriented rugosities in low latitudes.

An evolutionary classification should be interpreted as being consistent with phylogeny. This point has been stressed by Simpson in various works and is repeated here perhaps to the point of redundancy. And one of the main problems of morphologic classification based on phylogeny is the selection of characters which are homologous or parallel in nature, not convergent, since homology is always valid evidence of affinity.

The key to understanding the evolution of microfossil groups lies in an understanding of their basic biology which can be gained from studies in both the laboratory and field and on living and dead specimens. An understanding of the conditions governing growth in microfossils is of paramount importance in the micropaleontologist's attempts at paleoecologic reconstructing. Recent success in culturing planktonic microorganisms under laboratory conditions argue well for future studies in this area. Morphologic studies suggest that phenotypic variation is related to environmental factors and that taxonomy is a function of test ultrastructure and biomineralization. We may expect morphologic studies on high-latitude planktonic forms so that they may be utilized with greater precision in paleoclimatic studies. The general potential of morphological research in paleoecology-paleooceanography is high and this generally unexplored field of research is a fertile area for creative ideas and new techniques in the years ahead.

Plankton evolution

The abundance of planktonic microfossils and the relative completeness of the stratigraphic record in the deep sea render these organisms ideal for evolutionary studies. Recent studies have been made on the factors governing temporal fluctuations in phyto- and zooplankton diversity, and rates of evolution, and extinctions. Species diversity and rates of evolution among calcareous plankton exhibit a positive correlation with the $^{18}O/^{16}O$ paleotemperature curve, suggesting that climate is one of the primary factors influencing the evolution of plankton (Berggren, 1969; Haq, 1973).

With recent refinements in biochronology it is now possible to make refined studies on phylogenetic trends within evolving lineages. Although these trends have usually been based upon morphogenetic changes in the external skeleton, i.e. are essentially a measure of morphologic character change, recent application of amino acid biochemistry is providing a promising area of research into the genetic relationships within and between microfossil groups (King and Hare, 1972a, b; King, 1975). This type of study should yield data allowing a more "natural" classification of taxa and thereby provide a more realistic basis upon which to conduct investigations on evolution.

Biochemical oceanography

The relationship between marine geochemistry (which until recently has been the sanctuary of the physical chemists) and micropaleontology represents a fertile area of research. Perhaps the most apparent common ground of these two fields lies in the area of carbonate dissolution. Carbonate dissolution encodes a significant message representing a potential basic environmental hazard in the form of excess CO_2 by the early part of the 21st century. Such excess CO_2 in the atmosphere should be eventually neutralized in the ocean by the production, sedimentation and remineralization of carbonate-producing planktonic organisms. Remineralization rates of biogenic carbonate, which controls the alkalinity of the ocean, depend upon residence time in the undersaturated water column and absolute dissolution rates; we may expect considerable research efforts in this field in the future.

ECOLOGY: PALEOECOLOGY

Ecology is the study of the relationships between organisms and their environment. Thus it deals with nearly all levels of organization of life, from the individual organism to the whole community of organisms living in an area to the effect of climate and geological processes on these organisms.

The marine ecologic system, the marine ecosystem, includes organisms and the factors making up the physico-chemical environment in a system similar in many ways to the organization of an individual organism. As in the case of the interrelationship between organs in a body, the arrangement of components in the marine ecosystem are not haphazard. There is a history of development, a particular spatial orientation (for example a water mass), a time factor in the operation of the system, and the involvment of specific energistic sequences, such as food chains (trophic resource regimes) or chemical cycles, etc. This organization could be summed up as four types of orderliness:

Evolutionary: Ecosystem components are products of organic evolution of individuals adapted or adjusted to that particular environment. Evolution produces biotic communities of coexistent species of plants and animals mutually adjusted (or adjusting) to each other and their milieu.

Spatial: The spatial arrangement of coexistent species in a system is determined by the **ecological niche** of each species, which simply is how and where each species "makes its living". A niche thus includes the physico-chemical habitat of a species as well as the adaptive strategy it employs to succeed in that habitat. Another means of defining spatial orderliness in an ecosystem is by **stratification**; for example marine communities are often arranged in vertical "layers", such as the microplankton in the photic zone of the ocean.

Temporal: Organisms in the marine ecosystem are not randomly active during one day, or one year. There is a periodicity, or seasonality which species obey and which often allows more species to inhabit an area as long as they have different or slightly overlapping activity periods. The seasonal blooms of phytoplankton groups vary slightly but sequentially to allow the large number of microplankton species to coexist throughout the year in the photic zone. Vertical migration of some plankton is tuned to the phases of the moon, and near-shore microbenthos is affected by tidal cycles. Reproductive seasonality is another example of temporal orderliness in the marine ecosystem.

Metabolic or trophic: This describes the patterns of energy and material transformation in the marine environment. The chemical cycling of nutrient or other elements in the ocean (and from the land) and through the biotic community proceeds in an orderly sequence so that few chemical elements are permanently lost from the community. The $CaCO_3$ in the tests of marine microorganisms is a major contributor to the CO_2 cycling in the oceans.

From this short discussion the immense task of studying the complex ecology of any marine group is apparent. Nevertheless, using information on present ecologic relationships it is possible to examine fossils and interpret their relationship to paleoecosystems. The study of fossils and their paleoenvironment is called paleoecology; likewise the study of past distributions and geographic relations is termed paleobiogeography, which in one sense is really the charting of the distributions of ancient environments and their components. Paleooceanography, then, is really the study of the components of ancient marine ecosystems.

Until recently the principle of uniformitarianism has been the keystone of all paleoecologic interpretation. This principle states that processes and relationships operative in modern systems can be extrapolated to interpret analogous systems in the geologic past. Recent advances in benthic paleoecology, such as the studies of depth distributions of benthic ostracodes and foraminifera, have demonstrated that the spatial (depth) distributions of ancient benthic species are, in fact, not identical to the depth distributions of their Recent analogues. Such studies point out several caveats in paleoecologic studies worth mentioning:

(1) The more specific the features studied, the less likely they will be strictly analogous in past systems.

(2) The organism–environment cause-and--effect relationships may be misleading, causing one to doubt the principle of uniformitarianism, when in fact, the proper "cause" has not yet been found.

(3) The modern ocean is a geologically re-

cent feature, particularly in its temperature structure; hence, comparisons between pre-Miocene and Recent systems become tenuous if the "real" values of variable, for example temperature or salinity, in the past are not measured, but only inferred by analogy.

Despite these problems, the field of marine paleoecology or paleooceanography, is one of the fastest growing and creatively approached aspects of marine micropaleontology today.

Sea-floor distribution — paleobiogeography

Oceanic circulation is dependent upon dynamically interrelated aspects of geography and climate. The concept that the earth is made up of a number of interlocking plates whose geometry has been subject to cyclic geographic rearrangements through time is the central hypothesis of plate tectonics — an idea which has virtually revolutionized the study of earth history. Although the exact timing of some of these interrelated events (particularly the earlier ones) during the most recent re-arrangement which began almost 200 m.y. ago is not always known, the relative sequence and general relationships are adequately known.

Reduced to its simplest scenario the earth appears to have consisted essentially of a single "supercontinent", **Pangaea**, surrounded by "superocean" **Panthalassa** in the late Paleozoic. In the early Mesozoic two major components of this single land mass, **Gondwanaland** (in the south) and **Laurasia** (in the north) were separated by a triangular reentrant of Panthalassa, the **Tethys Sea**. Northward rifting of fragment(s) from northern Australia which later became part(s) of the Asian continent began around 180–160 Ma, approximately simultaneously with the initial opening of the central Atlantic by the rifting of the North American and African continents. The subsequent fragmentation of Gondwanaland (separation of India from western Australia) and opening of the Indian Ocean) began in the late Jurassic–early Cretaceous (ca. 130 Ma) approximately simultaneously with the opening of the South Atlantic. These events herald the beginning of the evolution of the present-day oceans and signal the gradual diminution of Tethys, and its ultimate evolution into the Mediterranean Sea — a process which was to take over 100 m.y., and was completed only recently within the life span of Man.

The transformation of biogeographic distribution patterns into fossil distribution (taphogeographic) patterns is a function of differential transport and preservation. Differential dissolution affects the preservation of planktonic organisms and has significance for biostratigraphy and paleooceanography. One of the major research areas at present is the role of selective dissolution on preservation of calcium carbonate and siliceous skeletons of microfossils (Berger, 1974). Recent studies have shown that coccolith ooze on the ocean floor and the well-preserved suspended coccoliths in the undersaturated water column are the result of accelerated sinking of coccospheres and coccoliths in the fecal pellets of small zooplankton (Honjo, 1976). Thus, the community structure of the euphotic layer is replicated with high fidelity in the depositional thanatocoenose. Destructive and non-destructive predation may be expected to play an important role in determining the nature of sedimentation patterns in the deep sea. The transfer of nutrient matter to the deep sea is also dependent upon fecal transport. Since nutrient supply is one of the major factors involved in controlling fertility in the upper layers of the ocean and thus sedimentation patterns on the bottom, we may expect to see an enhanced interest in this type of study with a view to elucidating patterns of paleofertility.

Taphogeographic sea-floor distribution studies are being conducted in most of the major planktonic groups, particularly the foraminifera, radiolaria and coccoliths, and these form the basic data by which many of the major paleogeographic, paleooceanographic-paleobiogeographic events have been dated and the history of regional and global oceanic circulation patterns have been delineated (Atlantic: Berggren and Hollister, 1974; Antarctic: Kennett and others 1975; global: Berggren and Hollister, 1977). Distinct latitudinal control on species diversity is seen within calcareous plankton whereas siliceous plankton exhibit a bimodal diversity maximum pattern, in the equatorial Pacific regions and again in polar regions.

Biogeographic distribution patterns of early Cenozoic calcareous nannoplankton and planktonic foraminiferal assemblages have been delineated for the North and South Atlantic (Haq and others, 1977). The latitudinal distribution through time allows recognition of certain assemblages which can be used as environmental indicators. On the assumptions that the latitudinal differentiation of early Cenozoic calcareous nannoplankton are related to a latitudinal temperature gradient and that the ecologic preferences of these assemblages are relatively stable through time, the latitudinal migrations which were recognized have been interpreted to have been caused by paleotemperature changes. Paleotemperature trends delineated by biogeographic migration patterns in Paleogene calcareous plankton are being correlated with paleotemperature records derived from the oxygen isotopes of marine Paleogene planktonic species, as well as trends depicted by some workers on the basis of terrestrial floras.

The role of post-depositional alteration of microfossils and the formation of deep-sea limestones and cherts through diagenetic changes is being intensively studied (Packham and Van der Lingen, 1973; Schlanger and Douglas, 1974). The diagenetic alteration of deep-sea sediments may lead to a variety of preservational states which, in turn, form the basis of acoustic stratigraphy as used by the geophysicists in mapping deep-sea reflector horizons. This forms but one example of the way in which micropaleontological and geophysical studies complement each other.

Micropaleontologic research in paleoceanography perhaps best illustrates the importance of the marriage between paleontology and geophysics. The microfossils are important as environmental indicators and as biostratigraphic tools. But the interpretation of past distribution patterns depends ultimately upon realistic paleogeographic and paleobathymetric reconstructions. The methodology exists for these reconstructions and the biostratigraphic and paleoenvironmental evidence of the microfossils places temporal constraints upon the spatial reconstructions of the geophysicist within the framework of plate tectonics. It is now possible to elucidate the history of circulation and sedimentation patterns in the ocean based on distributional patterns in various microfossil groups.

FUTURE TRENDS IN MARINE MICROPALEONTOLOGY — TOWARDS THE 21ST CENTURY

Marine micropaleontology has undergone a series of dramatic changes over the past decade, a trend that may be expected to continue into the foreseeable future. These changes have been primarily in the areas of biostratigraphy and biochronology, as well as in shifts in research emphasis to such areas as paleooceanography (including paleobiogeography and paleoclimatology) and plankton evolution. These changes have been due primarily to recent advances in technology, such as deep-sea coring, computers, and scanning electron microscopy, among others. As a result marine micropaleontology now plays a fundamental role in the interpretation of the history of the marine biosphere as may be seen from its ever increasing integration with marine geophysics and geology.

A comprehensive review of current research trends in marine micropaleontology is beyond the scope of an introductory textbook. At the same time we believe that the fact that micropaleontology is currently undergoing a "revolution" and "rejuvenation" warrants a brief overview of some of the more important trends of current research in this field. A more comprehensive, up-to-date summary (Berger and Roth, 1975) of current research trends will be found cited in the bibliography.

As the present is considered a key to the past by geologists, so too it holds the promise of the future. Where will marine micropaleontology be at the turn of the century? What are the major trends which will characterize its continued development during the last quarter of this century? It would seem that we can expect the major trends to lie within those areas of research which are at present being most actively pursued and which have been identified and briefly described above, as well as within several new areas which are only now in their infancy.

In the area of biostratigraphy and biochronology we foresee continued efforts at extending and refining planktonic biostratigraphic zonations (particularly in high latitudes), correlating them with zones in shallow-water areas based on large foraminifera, and the correlation of biostratigraphy and magnetostratigraphy and their calibration to radiochronology. Advances in this area should come primarily in the Mesozoic and early Cenozoic intervals. These studies will provide an increasingly reliable biochronologic framework within which global geologic and biologic processes may be delineated.

In the field of paleooceanography we include the diverse, but interrelated, areas of paleobiogeography and paleoclimatology. In the field of Quaternary paleoclimatology we may expect continued advances in understanding of the temperature and oceanic circulation over the last 1.6 m.y. as a clue to predicting future climatic trends. In the field of pre-Quaternary paleooceanography the use of oxygen and carbon isotope analyses will add significantly to our understanding of the biogeographic, geochemical (including productivity), climatic and circulation history of the oceans. Studies on the fluctuations in the areal distribution and relative abundance patterns of siliceous and calcareous sediments will continue to yield information on the history of productivity, dissolution and the CCD in the oceans and result in global syntheses of sedimentation and erosion patterns. Here we foresee a trend towards interdisciplinary studies involving micropaleontology, sedimentology and geochemistry.

The application of micropaleontology to studies on the geologic history of continental margins and slopes may be expected to increase in relation to increased exploration in these areas for petroleum and other mineral resources. Concomitant with such studies will be the continued use of various microfossils, particularly benthic foraminifera, in reconstructing the paleocirculation history of deep oceanic water masses and the tectonic history of various depositional basins and the subsidence history of foundered continental fragments in the oceans.

In summary, we may expect that during the course of the last quarter of this century there will be greater emphasis on quantitative analysis of assemblages which will provide greater precision in the areas of biostratigraphy and paleoecology. At the same time improvements in transmission and scanning electron microscopy may be expected to yield further insight into skeletal ultrastructure. The *menage à trois* of marine biostratigraphy, geomagnetism and radiochronology should provide a continually improved geochronologic framework for studies in earth history. Finally, it would seem safe to predict that the increase in interdisciplinary studies of marine micropaleontologists with physical and chemical oceanographers will continue to place micropaleontology in the mainstream of paleo(oceanographic) research.

SUGGESTIONS FOR FURTHER READING

Berger, W.H. and Roth, P.H., 1975. Oceanic micropaleontology: progress and prospect. *Rev. Geophys. Space Phys.*, 13(3): 561–635. [Comprehensive survey of research in oceanic micropaleontology during the period 1970–1975.]

Berggren, W.A., 1971. Oceanographic micropaleontology. *EOS*, 52: 249–256. [A summary of research in oceanic microplaeontology between 1967 and 1970].

Eicher, D.L., 1968. *Geologic Time*. Prentice-Hall, Englewood Cliffs, N.J., 141 pp. [A concise treatment of the concept and growth of geologic time-scale and its application.]

Kummell, B. and Raup, D. (Editors), 1965. *Handbook of Paleontological Techniques*. Freeman and Co., San Francisco, Calif., 852 pp. [Standard reference on general procedures, techniques and bibliographies of use to (micro)paleontologists.]

Moore, R.C. et al., 1968. Developments, trends and outlooks in paleontology. *J. Paleontol.*, 42: 1327–1377. [A comprehensive survey of paleontology (including micropaleontology) until the year 1967.]

CITED REFERENCES

Berger, W., 1974. Deep-sea sedimentation. In: C.A. Burke and C.L. Drake (Editors), *The Geology of Continental Margins*. Springer-Verlag, Heidelberg, pp. 213–241.

Berggren, W.A., 1969. Rates of evolution in some Cenozoic planktonic foraminifera. *Micropaleontology*, 15(3): 351–365.

Berggren, W.A. and Hollister, C.D., 1974. Paleogeography, paleobiogeography and the history of circulation in the Atlantic Ocean. In: W.W. Hay (Editor), *Studies in Paleo-oceanography. Soc. Econ. Paleontol., Mineral., Spec. Publ.*, 20: 126–186.

Berggren, W.A. and Hollister, C.D., 1977. Plate tectonics and paleocirculation: commotion in the ocean. *Tectonophysics*, 38(1–2): 11–48.

Cox, A., 1969. Geomagnetic reversals. *Science*, 163 (3864): 237–245.

Haq, B., 1973. Transgressions, climatic change and the diversity of calcareous nannoplankton. *Mar. Geol.*, 15(1973): M25–M30.

Haq, B., Premoli-Silva, I. and Lohmann, G.P., 1977. Calcareous plankton paleobiogeographic evidence for major climatic fluctuations in the early Cenozoic Atlantic Ocean. *J. Geophys. Res.*, 82(27): 3861–3876.

Honjo, S., 1976. Coccoliths: production, transportation and sedimentation. *Mar. Micropaleontol.*, 1(1): 65–70.

Kennett, J.P. and others, 1975. Cenozoic paleooceanography in the southwest Pacific Ocean, Antarctic glaciation, and the development of the circum-Antarctic currents. In: J.P. Kennett, R.E. Houtz and others, *Initial Reports of the Deep-Sea Drilling Project, 19.* U.S. Government Printing Office, Washington, D.C., pp. 1155–1169.

King Jr., K., 1975. Amino acid composition of the silicified matrix in fossil polycystine radiolaria. *Micropaleontol.*, 21 (2): 215–226.

King Jr., K. and Hare, P.E., 1972a. Amino acid composition of planktonic foraminifera. A paleobiochemical approach to evolution. *Science*, 1975 (4029): 1461–1463.

King Jr., K., and Hare, P.E., 1972b. Amino acid composition of the tests as a taxonomic character for living and fossil planktonic foraminifera. *Micropaleontology*, 18(3): 285–293.

Mayr, E., 1966. *Animal Species and Evolution.* Belknap Press of Harvard University, Cambridge, Mass., 797 pp.

Packham, G.H. and Van der Lingen, G.J., 1973. Progressive carbonate diagenesis at Deep Sea Drilling Sites 206, 207, 208, and 210 in the Southwest Pacific and its relationship to sediment properties and seismic reflectors. In: J.E. Andrews, R.E. Burns and others, *Initial Reports of the Deep Sea Drilling Project*, 21. U.S. Government Printing Office, Washington, D.C., pp. 495–507.

Schlanger, S.O. and Douglas, R.G., 1974. The pelagic–ooze–chalk–limestone transition and its implication for marine stratigraphy. In: K.J. Hsü and H. Jenkyns (Editors), *Pelagic Sediments: On Land and Under the Sea. Spec. Publ. Int. Assoc. Sedimentol.*, 1: 117–148.

Simpson, G.G., 1940. Types in modern taxonomy. *Am. J. Sci.*, 238: 413–431.

Simpson, G.G., 1953. *The Major Features of Evolution.* Columbia University Press, New York., N.Y., 434 pp.

Simpson, G.G., 1961. *Principles of Animal Taxonomy.* Columbia University Press, New York, N.Y., 247 pp.

Van Hinte, J.E., 1969. The nature of biostratigraphic zones. *Proc. First Int. Planktonic Conf., Geneva*, 2: 267–272.

CALCAREOUS MICROFOSSILS

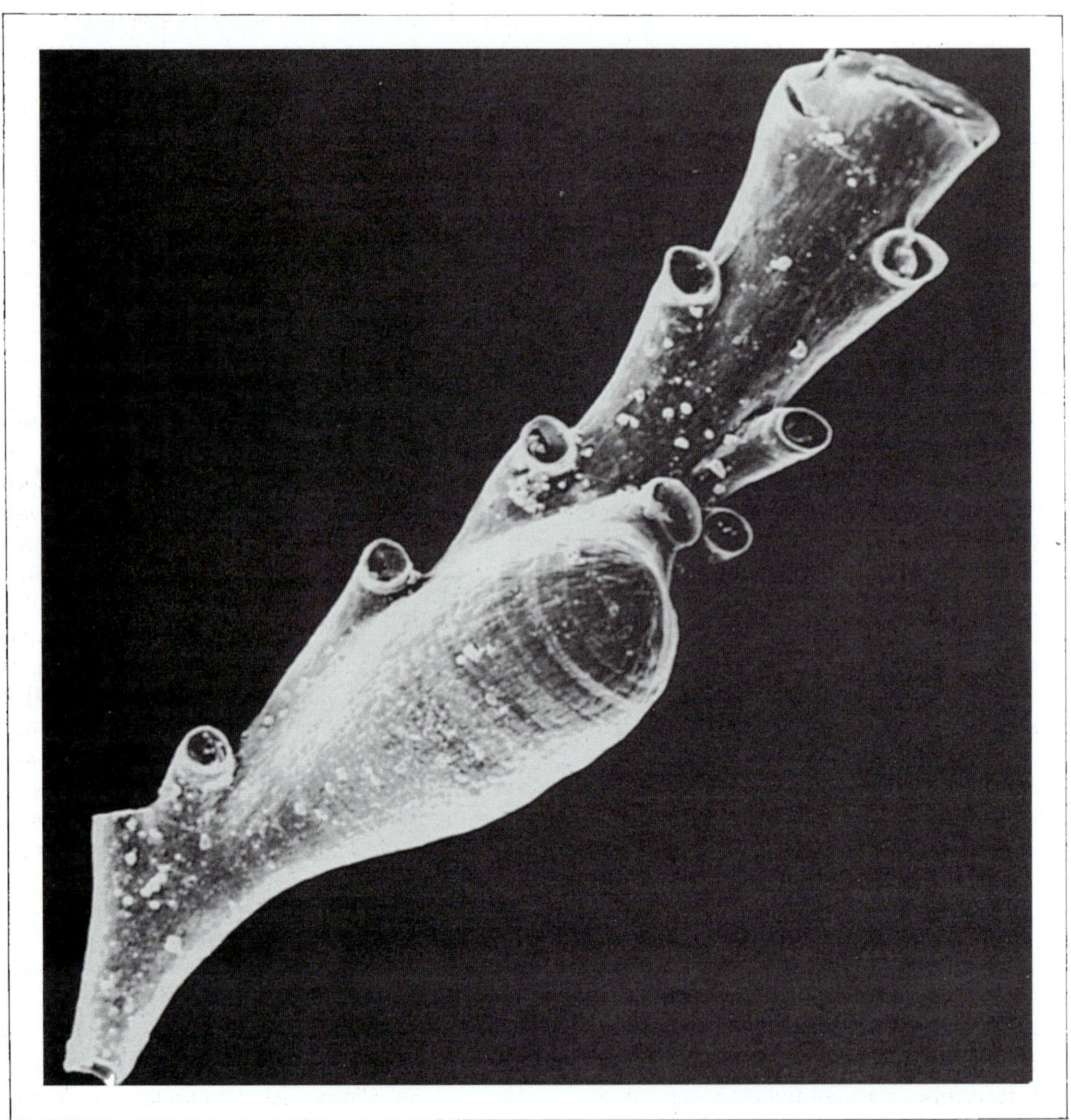

Segment of the cyclostomatous bryozoan *Crisia*

2

FORAMINIFERA

ANNE BOERSMA

INTRODUCTION

A date of 400,000 years B.P. is assigned to a Pleistocene sediment core bottom on the basis of the amino acid content in calcareous microfossils called foraminifera. The age of the opening of the North Atlantic is computed to be close to 120 m.y. B.P. from evidence lent by foraminifera. The temperature of the ocean 50 m.y. ago is estimated from the isotopes of oxygen in the shells of foraminifera. These and many other types of oceanographic information can be derived using the microscopic shells of this protozoan group.

Living both on the bottom and floating in the water column, these microorganisms presently inhabit the ocean from 5 to over 5,000 m depth. The bottom-dwelling forms have existed since Cambrian time and the planktonics since the Mesozoic. They are presently, along with the Ostracoda, the best known and most comprehensively studied of all the calcareous microfossils.

Foraminifera are roughly divided into three major groups: the planktonics, the smaller benthics, and the larger benthics which are distinguished by their larger size and by complex interiors which are visible only in thin section.

HISTORY OF FORAMINIFERAL RESEARCH

The study of any fossil group involves several steps. First the fossils are discovered and described, although their systematic position may not be determined until much later. As more descriptions accumulate, a system of organization gradually evolves. Within this system more and more taxa are described until a classification might be warranted based on new and improved information. The amassing of numbers of species descriptions then provides the groundwork for later interpretive studies. Foraminiferal research is presently at the stage where most cataloguing of species has been accomplished and researchers are in the process of interpreting this mass of data in light of recent theories of earth history.

Larger foraminifera of the genus *Nummulites*, which grow to a size of several millimeters, are abundant in the limestones used by Egyptians to build the pyramids at Gizeh, and their lenticular form led later travellers (5th to 1st century B.C.) to consider them as petrified lentils eaten by the slaves who built the pyramids.

Smaller foraminifera were first described and illustrated in the 16th and 17th centuries. Before the invention of the microscope in the late 1700's, early naturalists relied on ground magnifying lenses to view foraminiferal shells, some as small as 0.1–0.2 mm. Because foraminifera sometimes resemble the shells of gastropods and cephalopods in their coiled form and in their intricacy and beauty, early students mistook foraminiferal microfossils for tiny cephalopods or miniscule mollusks. Thus, in the 1700's foraminifera were still considered as fascinating oddities of nature, inorganic minutiae.

The French naturalist Alcide d'Orbigny revolutionized the study of foraminifera as well as several other branches of paleontology. In 1826, d'Orbigny described foraminifera as cephalopods and published an account of them in his principal work *Tableau Méthodique de la Classe des Céphalopodes*. Microcephalopods were divided into three classes, one of which we retain today as a designation for the whole group — Foraminifères, or

Foraminifera. D'Orbigny's lasting contribution was the first detailed classification of this group (1826).

After the monumental world cruise of H.M.S. *Challenger*, H.B. Brady illustrated the foraminifera dredged from the sea floor. His illustrations and a summation of the foraminiferal literature published up to that time appeared in the *Challenger* Reports in 1888. This publication has since been revised by Barker (1960) whose study remains the definitive work on Recent foraminifera of the world's oceans.

The second encyclopedic work on for a-minifera was produced by R.J. Schubert, who summarized all work up to 1921.

Several workers stand out as the initiators of modern foraminiferal studies. In the U.S.A. at the end of World War I, the most prolific student of the foraminifera was Joseph August Cushman. Cushman established a research laboratory in Sharon, Massachusetts, published voluminously, trained many students, wrote one of the most influential texts in the field (*The Classification and Economic Use of Foraminifera*), and established the first journal of foraminiferal studies, *Contributions of the Cushman Laboratory for Foraminiferal Research*. Cushman's key to the identification system has been followed up to the present time, with some revisions and improvements.

A major stimulus to foraminiferal research and one that was to change the direction and nature of foraminiferal studies was provided after World War I by the oil industry. In their intensive search for oil, they came to appreciate foraminifera as invaluable aids in the determination of the age as well as the depositional environment of strata. The impetus given by oil companies accelerated the study of foraminifera, in addition to many other microfossil groups.

Modern micropaleontology really began in the 1950's. Following the lead set by the oil companies the establishment of environmental indicator faunas was pioneered in the 1940's to 1950's by Fred Phleger and Orville Bandy. These indicator faunas remain today one of the primary means of interpreting the depositional environment of ancient sediments.

Modern biostratigraphy using foraminifera also blossomed during the 1950's. Before this time biostratigraphy of sediments was based on the stratigraphic ranges of benthic foraminifera. These ranges, however, often proved to be time-transgressive and hence were not always useful when correlating from one site to another. The first zonation scheme for the planktonic foraminifera was published by the Russian worker Subbotina for sections in the Caucasus Mountains. Then the seminal work by Hans Bolli (1957) on Tertiary sections from Trinidad provided the basis for most later low-latitude zonations. Since this time several workers have refined our biostratigraphic subdivisions, notably W.H. Blow and E. Pessagno in the 1960's and W.A. Berggren in the 1970's.

Applications of micropaleontological techniques to the ocean and to deep-sea cores also began during this period of the expansion of foraminiferal biostratigraphy and paleoecology. Schott (1935) and Cushman and Henbest (1942) published some of the first studies of foraminifera from deep-sea cores. Then Ericson and Wollin (1956, and later) published extensively on deep-sea cores, proposing the first extensive scheme for zonation of Atlantic deep-sea cores and for interpreting climatic change in these cores. Emiliani and Edwards (1953) showed the usefulness of oxygen isotopes in foraminiferal tests down deep-sea cores to recognize climatic changes and the base of the Pleistocene in deep-sea cores. Since the time of these early studies, deep-sea core studies have expanded enormously and were in part responsible for the birth of the Deep Sea Drilling Project which has recovered much longer deep-sea sediment cores by drilling with an oil rig. Enormous resources, both human and financial, are now going into the study of marine cores to interpret present and past oceanic environments and continental configurations.

The work of biologists towards clarifying the nature and systematic position of the foraminifera is significant. In the 1830's Felix du Jardin concluded that foraminifera were too primitive in their cellular makeup to be cephalopods or other such biologically complex organisms. Du Jardin proposed the name Rhizopoda based on his discovery that the protoplasm of foraminifera forms branches called pseudopodia, or rhizopods;

this name was used for nearly a century afterwards.

The majority of biological work on for a-minifera took place at the end of the 19th century, principally by the German and English biologists Rhumbler, Schaudin and Lister, who performed some of the first culture experiments with foraminifera and began the arduous study of foraminiferal life cycles. Their work has been continued in this century by Arnold, Jepps, Hedley, Myers, LeCalvez, and most recently by Lee and Bé, who has finally succeeded in culturing planktonic foraminifera.

BIOLOGY

Systematic position

Foraminifera belong to the Phylum Protozoa (Fig. 1). Unicellular, or acellular, organization is the single feature common to all the various members of this phylum, which is further divided into classes on the basis of the type of locomotor apparati present. Since foraminifera are non-flagellate, but possess flowing protoplasmic extensions termed **pseudopodia**, they are placed in the Class Sarcodina which includes the simplest of the Protozoa with respect to their cellular organization and specialization; but shelled sarcodines create skeletons of incredible beauty and structural complexity. (The Radiolaria, discussed in Chapter 9, are also skeletal-secreting sarcodines.)

Foraminifera are distinguishable from other Sarcodina by the possession of mineralized shells (although there are a few foraminifera that do not construct a shell but are covered

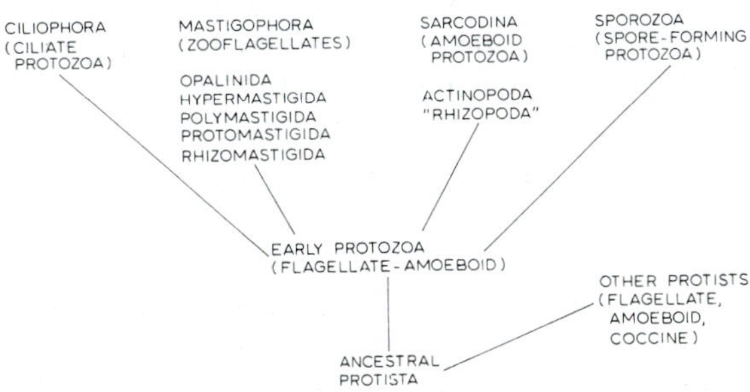

Fig. 1. Possible interrelations among protozoan groups showing the systematic position of the Rhizopoda, which include the Foraminifera.

by an organic material). In this chapter we will deal only with those forms that secrete or agglutinate a mineralized covering, since only these are preserved as fossils.

The cell and its contents

The cell of a foraminifer is a protoplasmic mass bounded by a limiting membrane. Some of the cell protoplasm is encapsulated in a secreted or agglutinated covering, the **test**, which may consist of one or more cavities termed **chambers**. When there is more than one chamber, a wall dividing one chamber from the next is termed a **septum**. Protoplasm is nevertheless continuous between chambers through a hole in the septum, the **foramen**, from which the name foraminifera (Latin *foramen* = hole; *ferre*, to bear) is derived. The protoplasm extends outside the test through the opening, or **aperture**, and surrounds the test in a mass of branching, anastomosing pseudopodia (Fig. 2).

Fig. 3 illustrates the contents of the protoplasm of a planktonic foraminifer. Like the Radiolaria, foraminifera may be uni- or multinucleate. The nucleus is generally round and consists of a nuclear membrane, nuclear sap, chromosomes, and the nucleoli which are the RNA-containing bodies. Although nuclei vary in size and number with different species, typically, single nuclei are larger than multiple nucleii.

The cell includes several typical protozoan structures, termed organelles. They consist of: the **Golgi bodies**, which play a role in cell secretions: **mitochondria**, sites of respiration; **ribosomes**, which contain RNA and are the sort of factory of protein synthesis; and **vacuoles**, which are fluid- or gas-filled droplets.

The protoplasm is generally colorless throughout but contains small amounts of organic pigments, iron compounds, brown and red deposits of fatty material, brown excremental particles, and green splotches produced by the presence of symbiotic algae living within the foraminiferal test. Many of these colorations can be detected in recently living specimens with a simple light micro-scope.

Foraminifera accomplish the life-sustaining and perpetuating activities of nutrition,

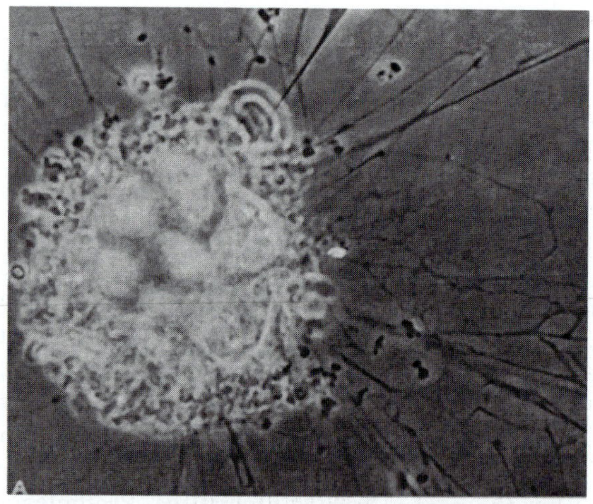

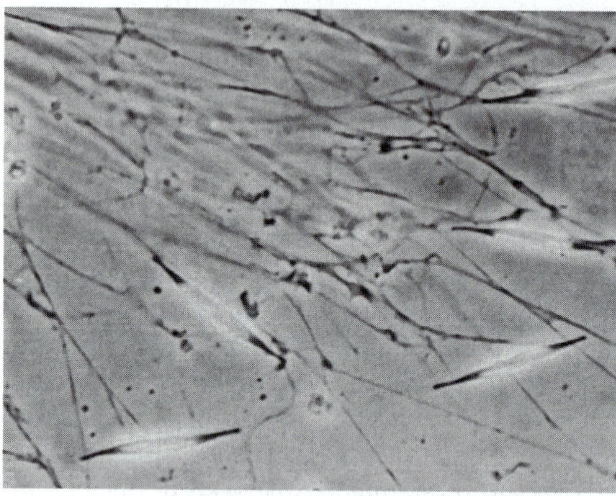

Fig. 2. The extant smaller benthic genus *Rosalina* in culture: A. The test is surrounded by food particles and debris, the pseudopodia radiate out from the test and can be seen to anastomose. × 350. B. Enlargement of the pseudopodia, showing diatoms and bacteria (dots) trapped in the pseudopodial network; these were later ingested by the foraminifer. × 420. (Courtesy D. Schnitker.)

motion, regeneration, respiration, reproduction, growth and test construction, all within the confines of this tiny protoplasmic mass.

The living animal

Nutrition

In a classic paper on protozoan motion Jahn and Rinaldi (1959) observed that pseudopodia are composed of thin filaments of a gel-like substance bent back on themselves somewhat in the manner of a conveyor belt. Like a conveyor belt, pseudopodia then move particles into and out of the inner protoplasm. This conveyor-like motion is termed **streaming** and is probably the most characteristic feature of foraminiferal protoplasm. Constantly streaming, the granular protoplasm issues from the aperture of the test; it may quickly withdraw into the test by this same streaming motion which resembles the movement of amoebas as they change their shape. The usefulness of streaming is particularly clear if we consider one activity of a foraminifer, food-gathering and excretion. Food is contacted and generally absorbed at the surfaces of the extended pseudopodia, chemically broken down into utilizable compounds, and streamed into the endoplasm. Similarly, material to be excreted agglomerates into small brown particles, is streamed out of the endoplasm through the pseudopodia, and is released into the water or dropped onto the substrate in trail-like fashion. In culture foraminifera feed several times a day, generally at the last chamber. Ingestion of food occurs outside the test except in some planktonic forms and some members of the benthic Miliolidae which have large apertures through which the food is drawn directly into the test. Several species of both planktonic and benthic foraminifera co-

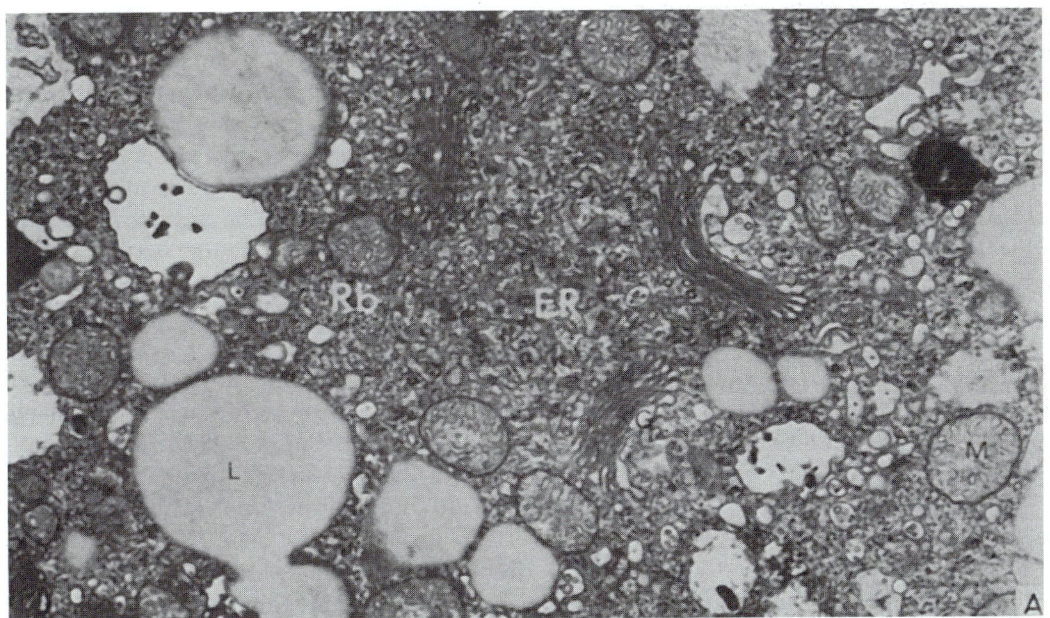

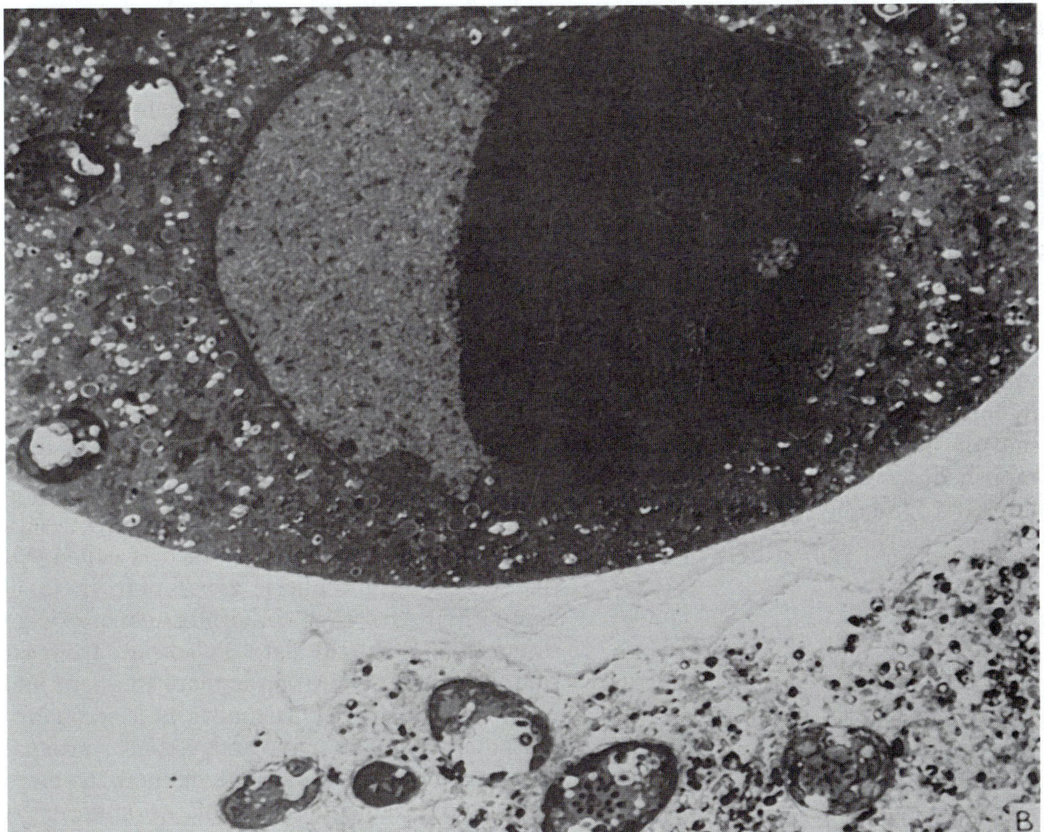

Fig. 3. A cross-section of the protoplasm of the planktonic genus *Globigerinoides*, showing organelles and other microstructure. A. Three lunate Golgi bodies (*G*) are located around a network of endoplasmic reticulum (*ER*) which in turn is surrounded by tiny granular (black dots) ribosomes (*Rb*). Mitochondria (*M*) and lipid bodies (*L*) of stored fats are also present in this area of the protoplasm which is considered to be an active site of secretion. × 2,040. B. The nucleus is contained in this second to last chamber of the test; this a single bi-lobed nucleus involved in the normal vegetative functioning of the test; smaller nuclei, involved in reproduction, are not present here. Note the more convolute membrane of the right lobe. × 2,240. (Courtesy Alan Bé).

habit with symbiotic algae, which by their photosynthesis are thought to provide a source of nutrition for the foraminifer.

Some benthic foraminifera feed by filter-feeding (*Bathysiphon*), but most are deposit-feeders or grazers, and at least one genus (*Entosolenia*) is known to be parasitic. In culture benthic foraminifera feed on diatoms, algae, bacteria, and particulate organic matter, as well as capture food particles from the protoplasm of a host. Planktonic foraminifera apparently ingest planktonic diatoms and other algae, silicoflagellates, copepods, and other microplankton. Foraminifera in turn are preyed upon by microscopic gastropods, pelecypods, pteropods, crustaceans, worms, and perhaps zooplankton as large as copepods. In many cases the foraminifera are merely part of the bulk sediment another benthic invertebrate is ingesting as it grazes the ocean bottom.

Movement

Benthic foraminifera may be sessile (*Cibicidella*, Fig. 18B1) or vagile. They can move by means of their pseudopodia on ocean bottom sediment, on algal fronds, or other substrates. Indeed it is often difficult to keep track of them in culture dishes as they have the annoying habit of moving up the edges of these dishes and out. Their movement may average close to 1 cm per hour, a respectable rate for an organism of approximately 0.5 mm in length. Several species of Recent planktonic foraminifera migrate through the water column up into the surface zone. This motion is not accomplished by any sort of pseudopodial flapping, but probably by changes in the chemistry, such as the gas content, of the protoplasm.

Reproduction

One of the most puzzling aspects of the foraminifera is a life cycle which most closely resembles that of certain plants. It will facilitate understanding the different modes of reproduction if we first consider a generalized life cycle of a foraminifer (Fig. 4). A life cycle is termed heterophasic when it characteristically contains two different phases, or types of reproduction and maturation. In many plants an asexual phase, or generation, alternates with a sexual generation. Among some foraminifera there is this same alternation of an asexual followed by a sexual generation. As seen in Fig. 4 the two principal phases are termed **schizogony**, the asexual phase; and **gamogony**, the phase involving sexuality. There are variations on this general cycle, the most frequent of which is absence or suppression of the phase involving sexuality. The benthic genus *Rosalina*, for instance, reproduces only asexually in culture. Most researchers assume that sexuality is a secondary addition to the life cycle and that foraminifera were initially asexually reproducing organisms which sometime in the late Paleozoic acquired the ability to reproduce sexually.

Individuals resulting from schizogony characteristically have a larger initial chamber, the **proloculus**, than the schizonts resulting from gamogony. The young gamonts with the larger proloculus are termed the **macrospheric generation**, while the individuals with smaller proloculi are called the **microspheric generation**. Exceptionally the microspheric generation may have a proloculus equal to or larger in size than the macrospheric generation. We call this production of two different initial morphologies in the course of a bi-phasic life cycle, alternation of generations, and the two morphologically distinct tests are termed **dimorphs**.

Dimorphic pairs are found among smaller and larger benthic foraminifera, but have not been recognized in planktonic genera. The microspheric generation with the smaller proloculus is termed the B form, whereas the megalospheric phase is called the A form. There are genera (*Cibicides*, *Triloculina*, and *Elphidium*) in which a third generation (see Fig. 4D) commonly occurs, which is called A1, the second megalospheric generation. In fossil populations the first dimorphic genera were recognized from the late Paleozoic. In most samples the A generation is more frequent and further suggests that sexuality is a secondary reproductive pattern and that asexual reproduction was the original and now the more frequent reproductive mode of the majority of foraminiferal species.

Growth

Growth has been witnessed in only a few shallow-water benthic genera. Non-periodic growth is characteristic of the single-cham-

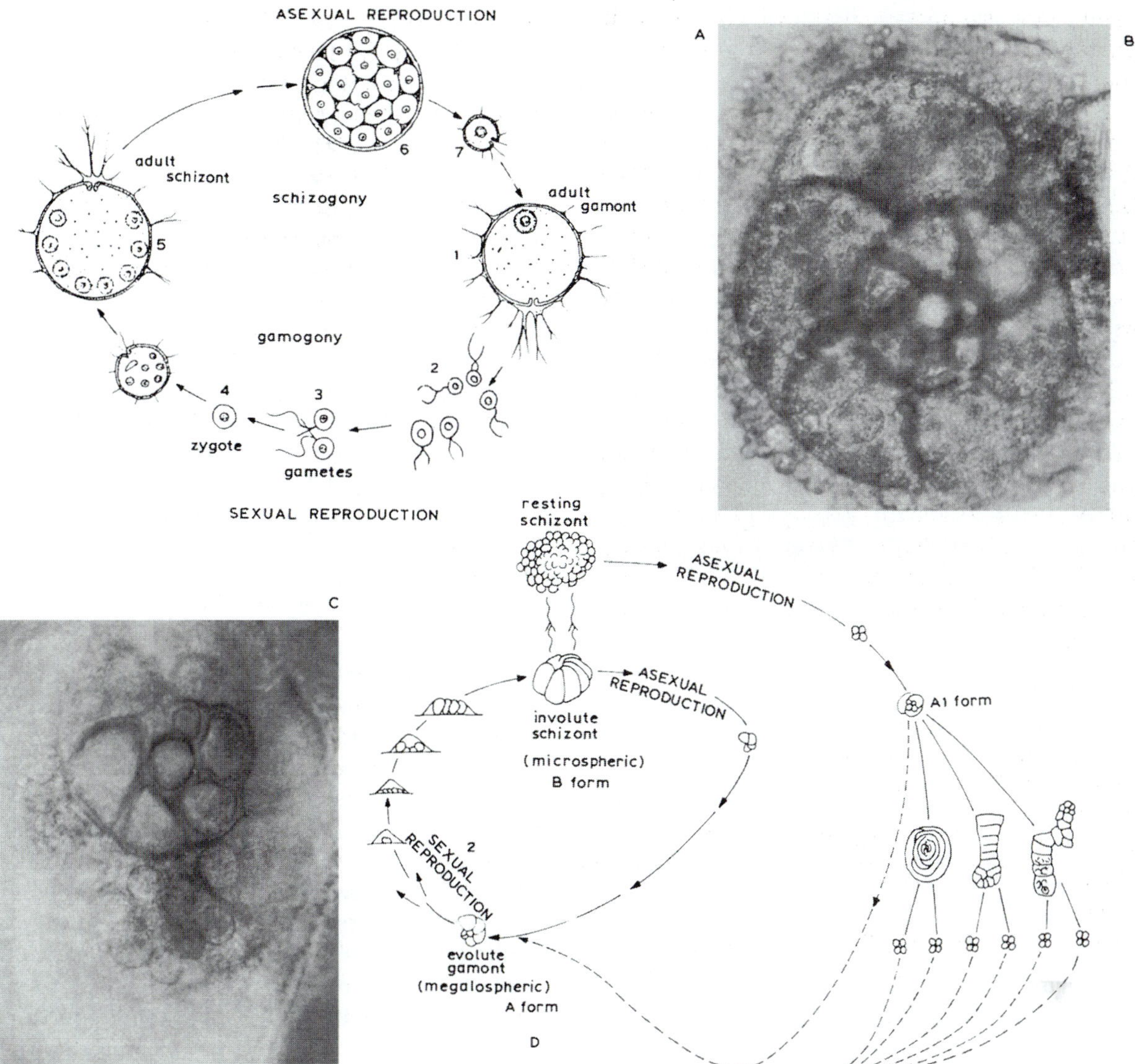

Fig. 4. Reproduction of foraminifera. A. Schematic drawing of a simple heterophasic reproduction cycle including the asexual phase (schizogony) and the phase involving sexuality (gamogony). Briefly, an adult (*1*) produces many bi-flagellate reproductive cells, gametes (*2*); when two join (*3*), the resulting individual, the zygote (*4*) then matures into an adult (*5*). Division of the protoplasm of this adult into several hundred offspring (*6*) is called schizogony. The emerging young (*7*) then grow into adults (*1*) and the cycle may be repeated. The round inclusions in the foraminiferal cells are the reproductive nucleii which divide during reproduction so that offspring receive some original material. B. Within the benthic genus *Rosalina*, reproducing in culture, can be seen the developing, well-defined offspring (lower left). × 400. C. Enlargement of the spiral region of the test (B) shows multiple offspring in the last whorl of the maternal test; this corresponds to phase *6* (schizogony) in the preceding diagram. × 420. (Courtesy D. Schnitker.) D. The production of several different morphologies during the reproductive cycle of the benthic genus *Cibicides* in culture; in this case, several different morphologies can be produced, and these multiple morphotypes can present problems to the taxonomist.

bered genera. The growth process generally takes place as depicted in Fig. 5. This is a case of periodic growth, as it involves a periodic increase in size by precipitation and accretion of mineral matter in the building of a new chamber. When a new chamber forms, the aperture of the last chamber generally becomes the foramen of the new chamber and permits connection of the protoplasm between the two chambers.

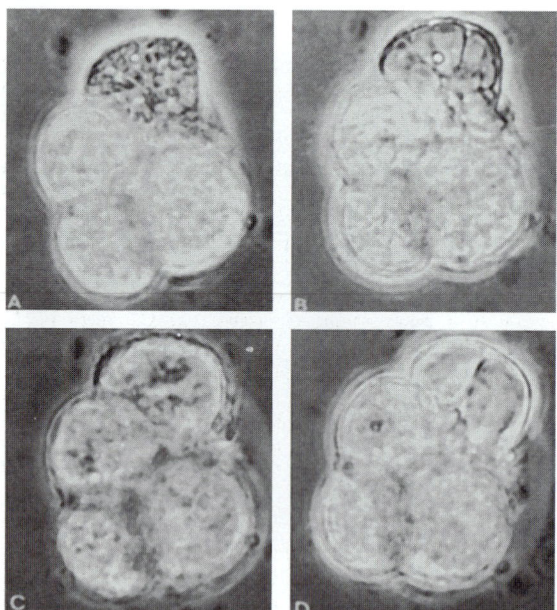

Fig. 5. Growth and chamber addition in a juvenile *Rosalina*. A. Juvenile individual, four hours old; naked cytoplasm has extended out of the test and now outlines the shape of the new chamber to be built. × 1440. B. One hour later, a membrane has formed at the surface of the cytoplasm. × 1440. C. One hour later, the calcification of the wall at the membrane has begun; the new chamber is partially filled with protoplasm. × 1360. D. Two hours later, calcification of the wall is nearly complete, and the new chamber is filled with protoplasm. × 1360. (Courtesy D. Schnitker.)

During chamber formation the foraminifer may encyst by gathering around itself a protective mass of sediment and/or the tests of other organisms. Foraminifera have been observed to encyst during chamber formation, feeding, mating, or as a resting body. Many living forms collected from the sea bottom come back to the lab still camouflaged in this mass of debris and protoplasm.

Test mineralization

We do not completely understand the process of test mineralization by the foraminiferal protoplasm. Test mineralization, at least in the major group of calcareous foraminifera, is apparently controlled by a proteinaceous organic template, a sort of anchor layer at the surface of the protoplasm which directs the growth of the calcite or aragonite crystallites. Attraction of the calcium ion from sea water by the amino acids in the protein template, is probably the key to mineralization, as the Ca ion then attracts the carbonate cation.

Although each planktonic and benthic foraminiferal species has a unique amino acid composition, the marked difference between amino acids in the calcitic and aragonitic species may help to explain the precipitation of aragonite in some foraminifera.

TEST MORPHOLOGY

Foraminifera are animals which build a shell; and for paleontologists the characteristics of the shell are the primary features which can be used to distinguish one species from another and hence to use these distinctions to form interpretations of time or environment.

Wall structure

The most readily obvious feature distinguishing one foraminifer from another is its wall type. Whether the foraminifer builds its test walls by cementing together exogenous grains, by carbonate mineralization, or by some combination of these two processes separates the three primary foraminiferal groups, the **agglutinated**, the **calcareous**, and the **microgranular** foraminifera. Original wall structure is considered to be a genetically stable feature of the test and hence to be relatively unaffected by environmental fluctuations.

Agglutinated wall structure

The geologically oldest method of test construction is the agglutination of particles together to form an external covering. Agglutinating foraminifers cement particles onto a layer of **tectin**, an organic compound composed of protein and polysaccharides. Extralocular protoplasm, calcite, silica, or ferruginous material are used to cement particles causing grains to be loosely bound in place or permanently cemented within this mineralized matrix. The grains involved may include assorted mineral particles such as sand grains, the tests of other microorganisms, distinctive sedimentary particles such as oolites, or microgranules of calcite scavenged off the deep-sea floor (see Fig. 6).

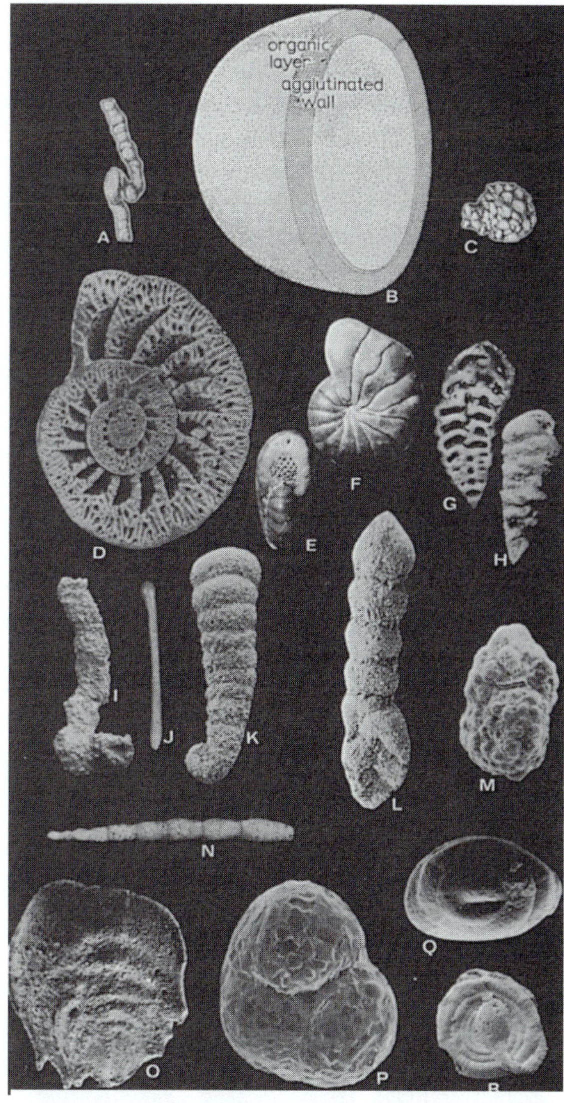

Fig. 6. Agglutinated foraminifera and wall structure: A. *Tolypammina* × 25). There are two primary groups of agglutinated foraminifera, the simple, non-septate forms belonging to the superfamily Ammodiscacea, and the more complexly coiled and septate forms in the superfamily Lituolacea; compare the simple structures of figures A, B, C, I, J, Q and R to the more complexly coiled forms in this figure. B. Schematic drawing of the structure of a simple agglutinated wall; a basal organic layer is overlain by particles which grade from finer inside to coarser at the outside. C. *Lagenammina* (× 19) from the early Cretaceous of the North Atlantic. (Courtesy H. Luterbacher.) D. Cross-section of the planispirally coiled *Cyclammina*, showing the spongy, convoluted infolding of the chamber walls, called labyrinthic structure; Recent, Gulf of Mexico. × 15. (Courtesy D. LeRoy.) E,F. End and front view, respectively, of *Cyclammina*, also from the Recent of the Gulf of Mexico. × 15. (Courtesy D. LeRoy) G. Cross-section through the wall of *Gaudryina* showing the simple wall structure typical of agglutinated foraminifera; both major superfamilies contain species with simple and complex interiors; Recent, Gulf of Mexico. × 15. (Courtesy D. LeRoy.) H. Whole specimen of the above. I. *Rhizammina* (× 50), early Cretaceous, North Atlantic. (Courtesy H. Luterbacher.) J. *Hyperammina* (× 15), Recent, Gulf of Mexico. (Courtesy D. LeRoy.) K. *Lituola* (× 15), Recent, Gulf of Mexico. (Courtesy D. LeRoy.) L. *Ammobaculites* (× 50), middle Cretaceous, South Atlantic. M. *Recurvoides* (× 15). This coarsely agglutinated test is composed largely of the smaller tests of foraminifera; Recent, Gulf of Mexico. (Courtesy D. LeRoy.) N. *Reophax* (× 15), Recent, Gulf of Mexico. (Courtesy D. LeRoy.) O. *Vulvulina* (× 70), late Eocene, Rio Grande Rise, South Atlantic. P. *Trochammina* (× 135), Recent, from the sub-Antarctic core RS12-5. (Courtesy R. Fillon.) Q. *Glomospira* (× 76), middle Cretaceous, South Atlantic. R. *Ammodiscus* (× 15), Recent, Gulf of Mexico. (Courtesy D. LeRoy.)

Some species will select grains of a specific size and composition to affix to the test; for example the genus *Psammosphaera* places a single elongate sponge spicule across the center of its test. Other agglutinated forms are non-selective and will employ any particle from a substrate as long as it lies in the appropriate size range.

In cross-section the wall of simple agglutinated forms is composed of a simple layer where particle sizes grade from finer on the inside to coarser on the outside (Figs. 6B, G). These walls may be pierced partially or wholly by tubules considered as pores.

In more complex agglutinated genera an outer smooth microgranular wall may cover a complex inner layer. This inner wall appears spongy and where portions of the inner wall are folded inward into the chamber of the test, the result is an intricate subdivision of the chamber interior. This type of wall structure has been termed **labyrinthic** (alveolar) in the agglutinated foraminifera (*Cyclammina*, Fig. 6D).

Microgranular walls

Microgranular walls evolved during the Paleozoic and are considered the link between the agglutinated and the precipitated tests in foraminifera. Microgranular particles of calcite cemented by a calcareous cement characterize this wall type and give it a sugary appearance.

Calcareous walls — hyaline type

Calcareous walls may be composed of either low- or high-Mg calcite, or aragonite which is confined to only two foraminiferal families. Hyaline calcareous tests are characterized by the possession of minute perforations in the test wall (Fig. 7A).

When the calcite or aragonite is arranged in prisms with their *c* axis normal to the test surface, the type of wall structure is termed **radial** hyaline. When crystallites are oriented randomly, the test is **granular** hyaline. Less common are hyaline tests composed of a single calcite crystal (*Carpenteria*).

In a cross-section of a hyaline wall (Fig. 7B), remnants of the organic layers are visible. These layers were organic and contained no visible structuring. Such layering occurs in all wall types. The second feature of some hyaline wall microstructure are the calcareous laminations, or **lamellae**. In non-lamellar foraminifera, a new chamber is accreted to the preceding chamber and separated from the previous chamber only by a single thickness of septum. By distinction, in lamellar groups such as the Rotaliidae, newly accreted material may form the new chamber as well as cover the entire outside of all previous chambers (Fig. 7). Thus lamellar tests are easily recognized by the thicker early chambers, the thinner later chambers and the single-thickness final chamber. This pattern varies, of course, when the foraminifer is enrolled.

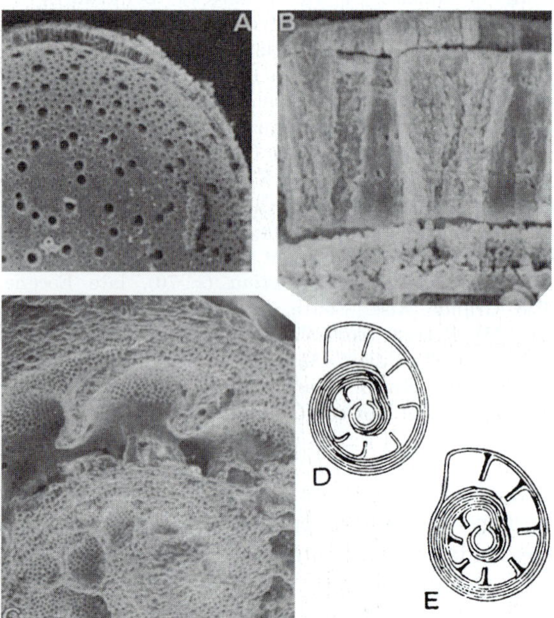

Fig. 7. Hyaline wall structure. A. Cross-section of the wall of the planktonic genus *Orbulina* (× 80). Circular holes are pores. Note the varying size and location of pores; these have been related to varying environmental conditions and can be used for environmental interpretations. B. Cross-section of the wall of the smaller benthic genus *Uvigerina*, showing the calcitic lamellae and the former position of the organic layers. × 750. C. Cross-section of the large benthic genus *Amphistegina*, showing the multi-lamellar outer wall and the characteristically multi-lamellar septae. × 105. (Courtesy R. Sherwood). D, E. Schematic drawing of the lamellar structure of foraminifera, showing how the sequential lamellae are added resulting in multiple lamellae on earlier chambers, and single-lamellar final chambers.

Planktonic foraminifera belong to the calcareous lamellar foraminifera and are further subdivided into two families on the basis of the surface texture of their walls. The smooth, non-spinose Globorotaliidae are thus distinguished from the spinose Globigerinidae (Fig. 8).

Calcareous walls — porcelaneous type

The term porcelaneous derives from the shiny, smooth appearance of the tests and is the result of the orientation of submicro-

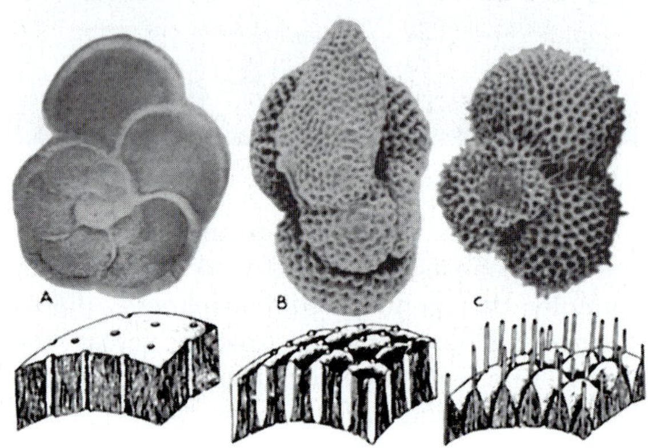

Fig. 8. Wall textures of the three main groups of planktonic foraminifera. A. Smooth-walled *Globorotalia* and associated genera are highly resistant to solution (see Fig. 25). B. The cratered wall of *Globigerinoides* is apparently most susceptible to solution, as these forms are the first to be lost from an association undergoing dissolution. C. *Globigerina* and associated forms in well-preserved samples will possess elongate spines, believed in some cases to be calcified cores of rhizopods; these forms show medium resistance to solution.

scopic crystallites of calcite that form the chamber walls. These crystallites may be randomly arranged or organized in a brick-like pattern, but both patterns give the test a smooth, opaque appearance in polarized light. Under crossed nicols the calcite crystallites

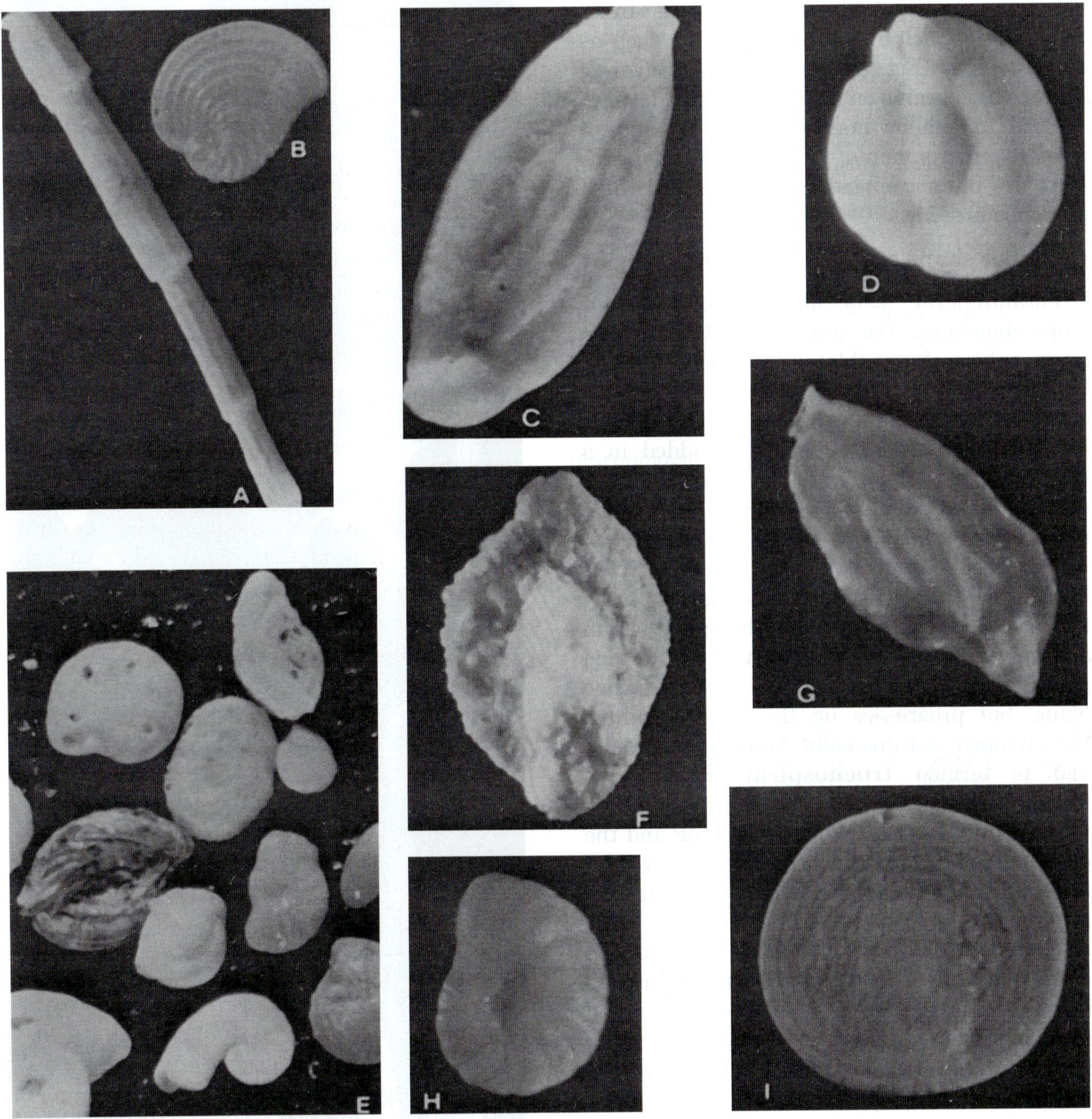

Fig. 9. Representative porcelaneous foraminifera (Miliolidae). A. *Articulina* (× 40), Recent Bahama Banks; miliolids are separated primarily on the basis of the type of coiling arrangement each possesses; thus, the milioline rectilinear *Articulina* belongs to a separate family from the solely milioline coiled *Triloculinella* (D) or the partially planispiral *Peneroplis* (B). Milioline coiling is also called streptospiral coiling, as successive coils are added in different planes; apertural type is also a feature important in distinguishing different miliolid groups. B. Peneroplis (× 10), Recent, Bahama Banks. C. Quinqueloculina. (× 40), Recent, Bahama Banks. D. *Triloculinella* (× 40), Recent, Bahama Banks. E. Miliolids in a carbonate beach sand from Puerto Rico. F. *Dentostomina* (× 10), recent, Bahama Banks. Notice the agglutinated outer covering composed of fine carbonate particles; on many forms this agglutinated covering is originally persent, but lost with transport and/or burial. G. *Quinqueloculina* (× 40), Recent, Bahama Banks. H. *Peneroplis* (× 12), Recent, Bahama Banks. I. *Sorites* (× 13) same as H. (Figs. A–D, F–I, Courtesy S. Streeter.)

pachyderma) and larger species in warmer water bodies or at low latitudes (*Pulleniatina*, *Globotruncana*). Diversity is lower at high latitudes and increases toward the equator. Such generalizations can easily be applied to Paleogene or Cretaceous planktonic foraminiferal faunas.

That oceanic circulation patterns affect the distribution of planktonic foraminifera is generally accepted; there are some local patchy distributions, however, which cannot be explained by this mechanism.

The relationship between planktonic foraminifera and trace element concentrations, nutrient elements, oxygen, turbulence, or turbidity is not known. Planktonic foraminifera live throughout the oceans, even in the ice in the Arctic Ocean. They do not, however, generally live in coastal waters less than 100 m and diversity drops markedly in waters less than 300 m. As some planktonic foraminifera contain photosynthetic symbionts, it is assumed that these forms are restricted to the photosynthetic zone in the water column. Several planktonic species (*Chiloguembilina*, *Heterohelix*) characterize the oxygen minimum zone of the water column and thus must prefer or tolerate low oxygen levels.

Distribution of living foraminifera

Foraminifera have been reported from marine environments extending from tide pools in a marsh to the abyssal plains. Each environment is characterized by its particular species, their diversity and densities. We consider that past environments may have contained many analogous components and hence modern environmental indicator faunas are carefully applied to the understanding of both Recent and past environments.

One curious feature of benthic foraminifera is the similarity of faunas in geographically widespread environments characterized by many similar chemical or physical parameters. And many benthic foraminifera have essentially cosmopolitan distributions, both in the Recent and in the past. Thus, it is possible to look at shelf environments around the world's oceans and find many of the same species in marshes of England and the northeastern United States. The endemism, common among the larger invertebrates, is relatively rare in foraminifera.

Carbonate platforms, reefs, and back reefs

The carbonate platform and reef environment is one of the geologically oldest and most complex of marine ecosystems. Modern reefs occur geographically between approximately 30°N and 30°S latitude, in areas with high light penetration, warm waters, and high dissolvedp calcium carbonate. They are characterized by waters with high salinities and turbulent conditions. Reefs are often adjacent to or on the edges of shallow-water platforms, such as the Bahama Banks.

Foraminifera occur in coral reef environments either as adherent forms (*Homotrema*, *Miniacina*) which may contribute to the construction of the reef framework, or as epifauna in niches developed within the reef framework (*Calcarina*, *Amphistegina*, *Marginopora*, Fig. 18).

Smaller benthic foraminifera are one of the primary contributors to the sediments of shallow carbonate platforms, second in importance only to the calcareous algae. They attach to sea weeds and grasses, algal and coral fragments; this foraminiferal epifauna may even occur on coral sands which are exposed to air during low tide. Larger foraminifera inhabit these shallow waters in association with the

Fig. 18. Typical foraminifera found on carbonate platforms and in proximity to reefs; morphologic analogous in ancient sediments indicate similar environments. A. Reef-associated foraminifera and their locations relative to the reef structure: *1*, *Amphistegina*; *2*, *Miogypsina*; *3*, *Peneroplis*; *4*, miliolids; *5*, *Alveolinella*; *6*, *Cycloclypeus*. B. Carbonate platform foraminifera from the Bahama Banks, Recent: *1*, *Cibicidella*; *2*, *Acervulina*; *3*, *Triloculina*; *4*, *Articulina*; *5*, platform sediments, including foraminiferal, algal, and pelecypod debris; *6*, *Dentostomina*; *7*, *Peneroplis*; *8*, *Dentostomina*; *9*, *Planorbulina*. (All photographs × 16. Courtesy S. Streeter.)

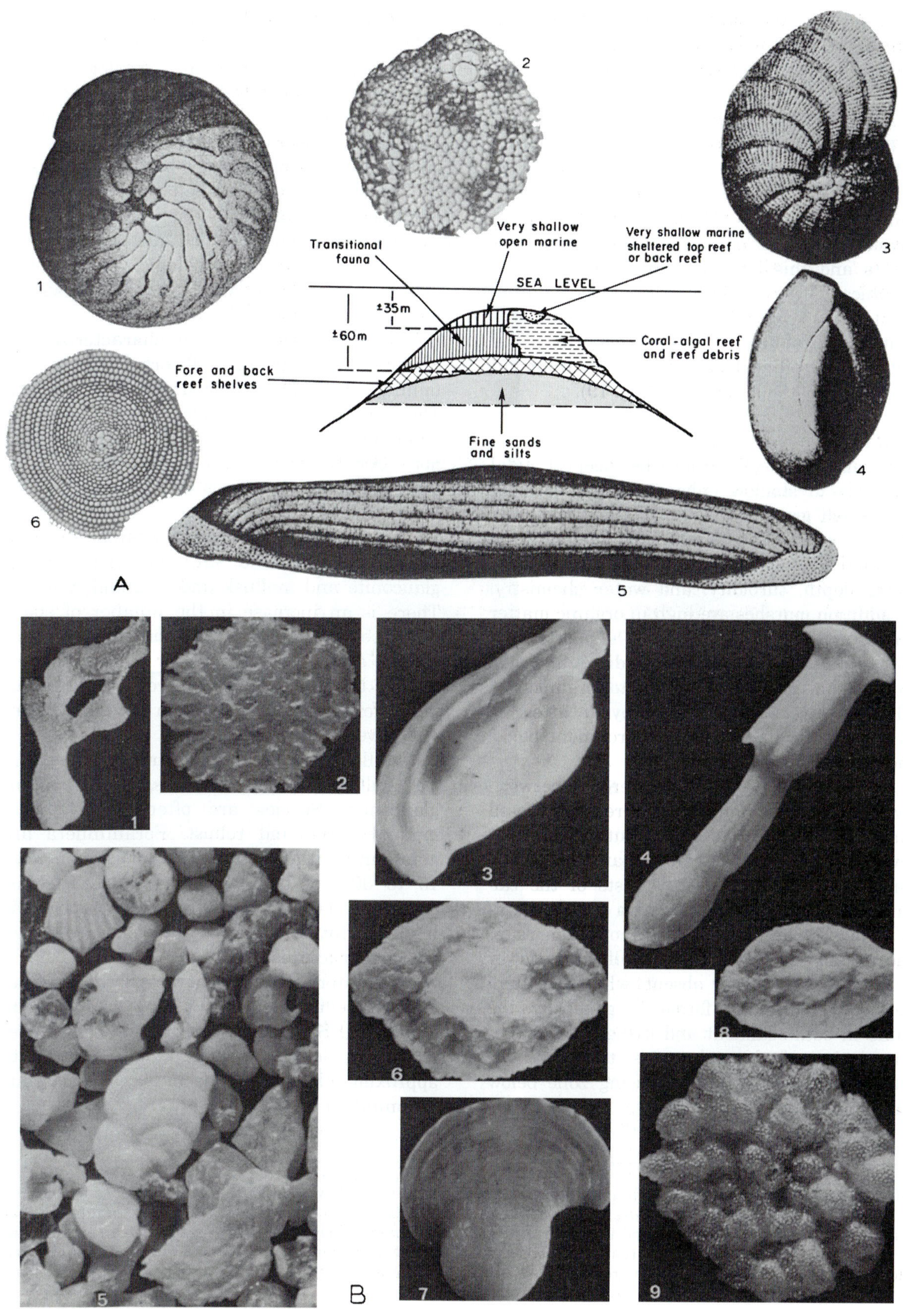
Very shallow open marine
Very shallow marine sheltered top reef or back reef
Transitional fauna
SEA LEVEL
±35m
±60m
Coral-algal reef and reef debris
Fore and back reef shelves
Fine sands and silts
A
B

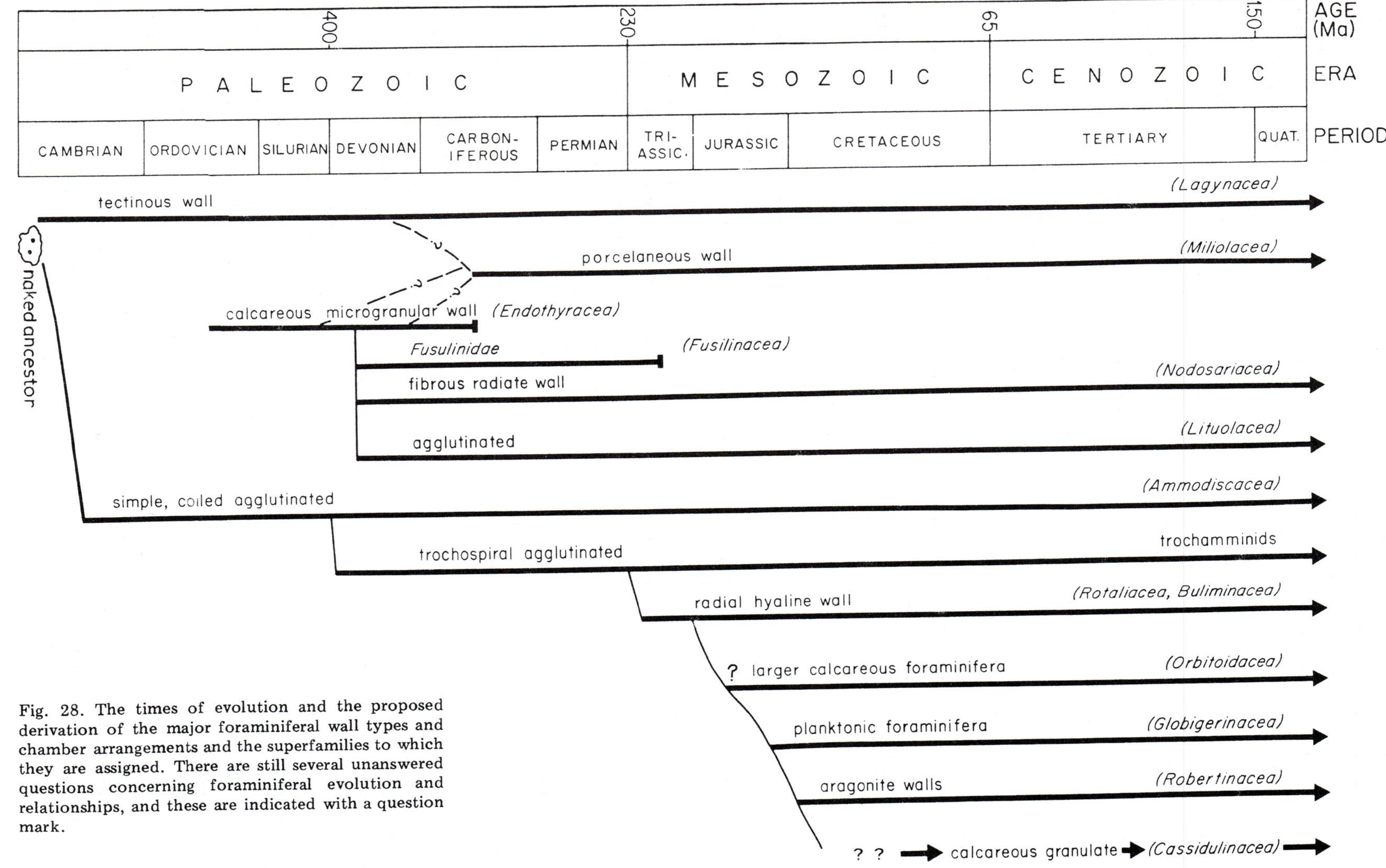

Fig. 28. The times of evolution and the proposed derivation of the major foraminiferal wall types and chamber arrangements and the superfamilies to which they are assigned. There are still several unanswered questions concerning foraminiferal evolution and relationships, and these are indicated with a question mark.

by the hypothesis that the oceans at this time finally became supersaturated with calcium carbonate and so foraminifera could use this compound to construct a test. Others suggest that microgranularity evolved as foraminifera advanced into carbonate-rich environments where they could easily use calcium carbonate as cement for sedimentary particles. Eventually, then, the foraminifera could precipitate their own calcium carbonate. The two new Devonian superfamilies with microgranular tests were the Parathuraminacea and the early Endothyracea.

Foraminiferal faunas of Devonian age have been retrieved from two facies, shallow carbonate facies and basinal facies. Each environment contains a unique fauna, although the microgranular genus *Parathurammina* apparently ranged through both environments. The Devonian is considered a period of extensive carbonate environments, as evidenced by the widespread limestones and dolomites of this age. It may be no coincidence that foraminifera first invaded and survived in carbonate environments of this time. At the end of the Devonian, tectonism along the Acadian and Uralian sutures produced environmental changes affecting the foraminifera.

By the termination of the Devonian the following families are represented in foraminiferal faunas: the Lituolidae, Nodosariidae, Endothyridae, and Parathuramminidae (all new families); as well as the preexisting Ammodiscidae, Astrorhizidae, Tournayellidae, Saccamminidae, Moravamminidae, Nodosinellidae, Colaniellidae, Ptychocoadiidae, and Semitextulariidae.

The Carboniferous was a period of radiation among pre-existing families of foraminifera, particularly the new microgranular families from the Devonian. The endothyrids, in particular, built tests of increasing complexity and size, giving rise during this time to the family Fusulinidae, probably the most important and complex of all Paleozoic foraminifera (Fig. 29). The increasingly complex interiors of the microgranular foraminifera were paralleled by the evolution of internal complexity (labyrinthic walls) among the agglutinated Lituolidae.

A significant evolutionary novelty in morphology at this time was the evolution of trochospiral coiling. Before this time there were planispiral involute (*Haplophragmoides*) or planispiral evolute (*Forschia*) species; then trochospiral coiling appeared in both microgranular (*Tetrataxis*) and agglutinated (*Trochammina*) genera. These calcareous trochoid forms are thought to be ancestral to the very important Mesozoic and Cenozoic rotaloid foraminifera.

During the Carboniferous foraminifera became more significant members of invertebrate communities and important rock-forming elements.

In the Mississippian there are limestones composed primarily of the genus *Endothyra*. Abundant fusulinid limestones occur in the late Pennsylvanian and Permian around the world. Such limestones apparently formed on shallow tropical shelves rich in calcium carbonate.

In the late Carboniferous-Permian the first porcelaneous family, the Agathamminidae, appeared. These forms are thought to possess an imperforate porcelaneous inner wall and a finely particulate, agglutinated outer wall. Some porcelaneous genera today also possess two outer walls built onto a tectinous basal layer.

The Permian could really be termed the period of the fusulinids as this group radiated into more than 5,000 species at this time. Less important, though well represented in the Permian, were the smaller foraminifera. During the late Paleozoic seventeen new families of smaller foraminifera evolved, but this radiation was abruptly terminated by a world-wide episode of mass extinctions which eliminated the fusulinids, the endothyrids, and members of ten other foraminiferal families. Both shallow and deep living macro-invertebrate taxa also became extinct at this time.

Beginning in the early Carboniferous the earth developed glacial conditions at the south pole and later at the north pole, as well. This glaciated earth must have been significantly different from the earlier Paleozoic in terms of climate, ocean circulation, sea levels, and available carbon. In addition, the late Paleozoic was a time of active tectonism and continental movements, all of which combined to alter drastically conditions in the hydrosphere to the detriment of marine ecosystems.

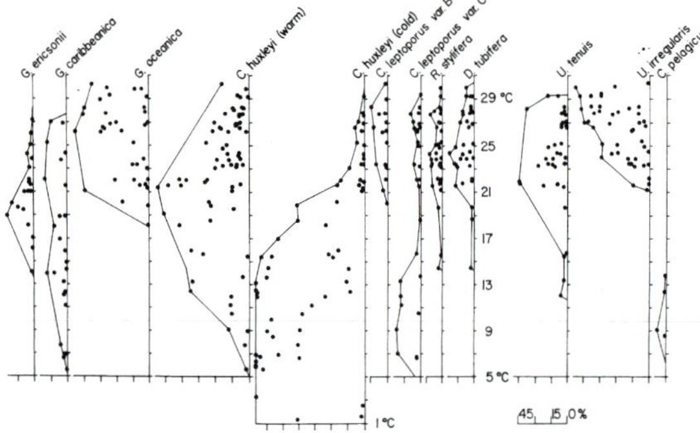

Fig. 7. Relative abundance of coccolithophore species in the Pacific Ocean plotted against mean surface temperature of water masses. Such data assign temperature ranges to taxa and enhance their paleoclimatic usefulness. (After McIntyre and others, 1970.)

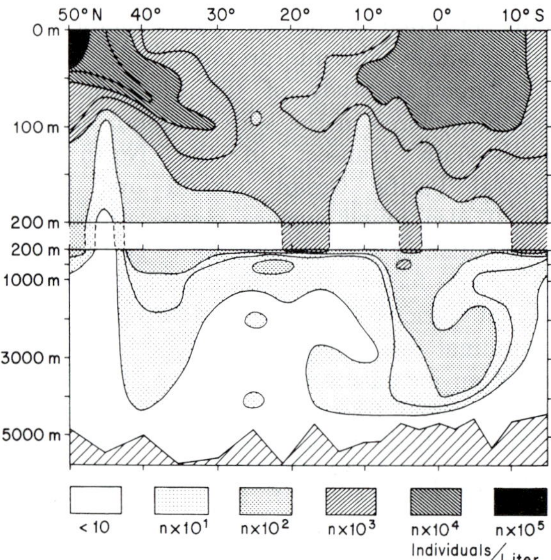

Fig. 8. Vertical distribution of coccolithophores along a north–south transect in the North Pacific. Note the wide variations in the number of individuals at various depths and high concentrations at 50°N and around the equator. (After Okada and Honjo, 1973.)

blages within the upper 200 m of the water column, which also changed with latitude. There was also a wide variation in the number of individuals with depth at various latitudes (Fig. 8).

MAJOR MORPHOLOGIC GROUPS

Calcareous nannoplankton constitutes a diverse group of morphological forms, many of which are either clearly related or show some similarity to the extant coccolithophores (forms with coccolith-like shields). Other forms with no clear morphologic relationship to coccolithophores (e.g. discoasters) occur as calcareous microfossils within the same size fraction as coccoliths and may form a substantial part of the nannofossil assemblages. Thus both the coccolithophores and the associated non-coccolithophore nannoliths are traditionally studied together by the nannopaleontologists.

Because of the very diverse nature and at times unknown affinities of the morphologic groups, no satisfactory classification for calcareous nannoplankton has been suggested to date. Also because of the dual, plant and animal, characteristics of nannoplankton (see section on biology), they are claimed by both the botanists and zoologists, and a complicated double systematics has developed over the years. Most present-day nannopaleontologists, however, favor the plant origin, and essentially by consensus, the Code of Botanical Nomenclature is used for the description of taxa. Botanists usually include coccolithophores in Order Heliolithae, Class Coccolithophyceae and Division Chrysophyta of the Plant Kingdom.

It is beyond the scope of this text to discuss the details of taxonomic and classificatory problems. Instead we have chosen to introduce the major groups of nannoliths that most commonly occur in nannofossil assemblages by grouping them under informal epithets. The groups are arranged into three categories: (a) those that show clear relationship to coccolithophores; (b) non-coccolithophores, but common nannoliths; and (c) common *incertae sedis* genera (for a summary of major groups, see Table I). With the exception of four groups (coccolithids, zygodiscids, braarudosphaerids and thoracosphaerids) that occur in both Mesozoic and Cenozoic sediments, all other morphologic groups are either restricted to the Mesozoic or the Cenozoic. For each group the most commonly occurring genera have been illustrated. See Fig. 9 for commonly used coccolith and discoaster terms. For details of terminology of other nannolith groups, the reader is referred to Farinacci (1971).

The catalogue of genera and species of nannofossils by Farinacci (1969 and later)

TABLE I

Key to major morphologic groups of calcareous nannoplankton

A. COCCOLITHOPHORES AND RELATED NANNOLITHS	
1. Arkangelskiellids	Family Arkhangelskiellaceae Bukry (Mesozoic)
2. Coccolithids	Family Coccolithaceae Poche (Mesozoic and Cenozoic)
	Family Prinsiaceae Hay and Mohler (Cenozoic)
	Family Helicosphaeraceae Black (Cenozoic)
3. Podorhabdids	Family Podorhabdaceae Noël (Mesozoic)
4. Pontosphaerids	Family Pontosphaeraceae Lemmermann (Cenozoic)
5. Rhabdosphaerids	Family Rhabdosphaeraceae Lemmermann (Cenozoic)
6. Stephanolithids	Family Stephanolithionaceae Black (Mesozoic)
7. Syracosphaerids	Family Syracosphaeraceae Lemmermann (Cenozoic)
8. Zygodiscids	Family Eiffellithaceae Reinhardt (Mesozoic)
	Family Zygodiscaceae Hay and Mohler (Cenozoic)
B. NON-COCCOLITHOPHORE NANNOLITHS	
1. Braarudosphaerids	Family Braarudosphaeraceae Deflandre (Mesozoic and Cenozoic)
2. Ceratolithids	Family Ceratolithaceae Norris (Cenozoic)
3. Discoasterids	Family Discoasteraceae Vekshina (Cenozoic)
4. Fasciculithids	Family Fasciculithaceae Hay and Mohler (Cenozoic)
5. Heliolithids	Family Heliolithaceae Hay and Mohler (Cenozoic)
6. Lithastrinids	Family Lithastrinaceae Thierstein (Mesozoic)
7. Lithostromationids	Family Lithostromationaceae Haq (Cenozoic)
8. Sphenolithids	Family Sphenolithaceae Deflandre (Cenozoic)
9. Thoracosphaerids	Family Thoracosphaeraceae Schiller (Mesozoic and Cenozoic)
C. GENERA *INCERTAE SEDIS*	
1. *Isthmolithus* (Cenozoic)	4. *Nannoconus* (Mesozoic)
2. *Microrhabdulus* (Mesozoic)	5. *Triquetrorhabdulus* (Cenozoic)
3. *Micula* (Mesozoic)	

and the annotated index and bibliography of Loeblich and Tappan (1966 and later) can also be consulted for description and validity of taxa.

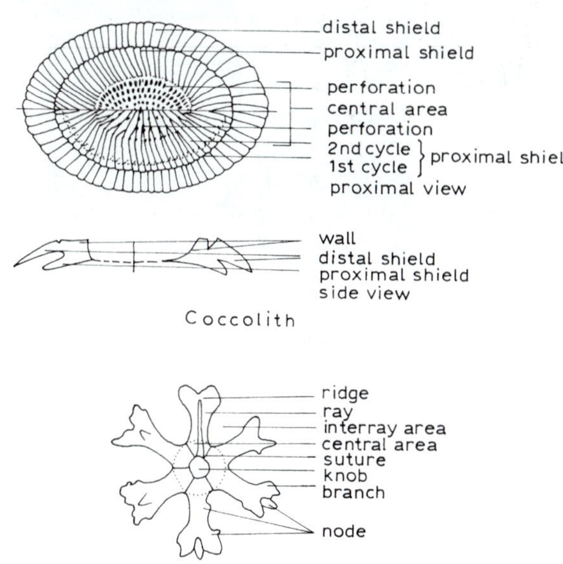

Fig. 9. Coccolith and discoaster terminology. (After Farinacci, 1971.)

Nannofossils can be studied rapidly for biostratigraphic analysis with the help of a light microscope at about × 1000. Smear-slides can be either made directly from raw sediment samples or from suspensions that have been cleaned chemically and short-centrifuged to concentrate nannolith-size particles. The use of normal, phase-contrast and cross-polarized light conditions are necessary to bring out finer details and the peculiar optical properties that help differentiate species. Transmission and scanning electron microscopes have been used increasingly in recent years to study the ultrastructure of nannoliths, the details of which are not discernible under light microscope. The reader is referred to Stradner and Papp (1961) and Hay (1965) for details of light microscopic methods and to Reinhardt (1972) for a discussion of the

TABLE III

Stratigraphic ranges and light microscopic illustrations of selected early Cenozoic nannofossil species (figures *a* under phase-contrast and *b* under cross-polarized light conditions; all figures × 3,200)

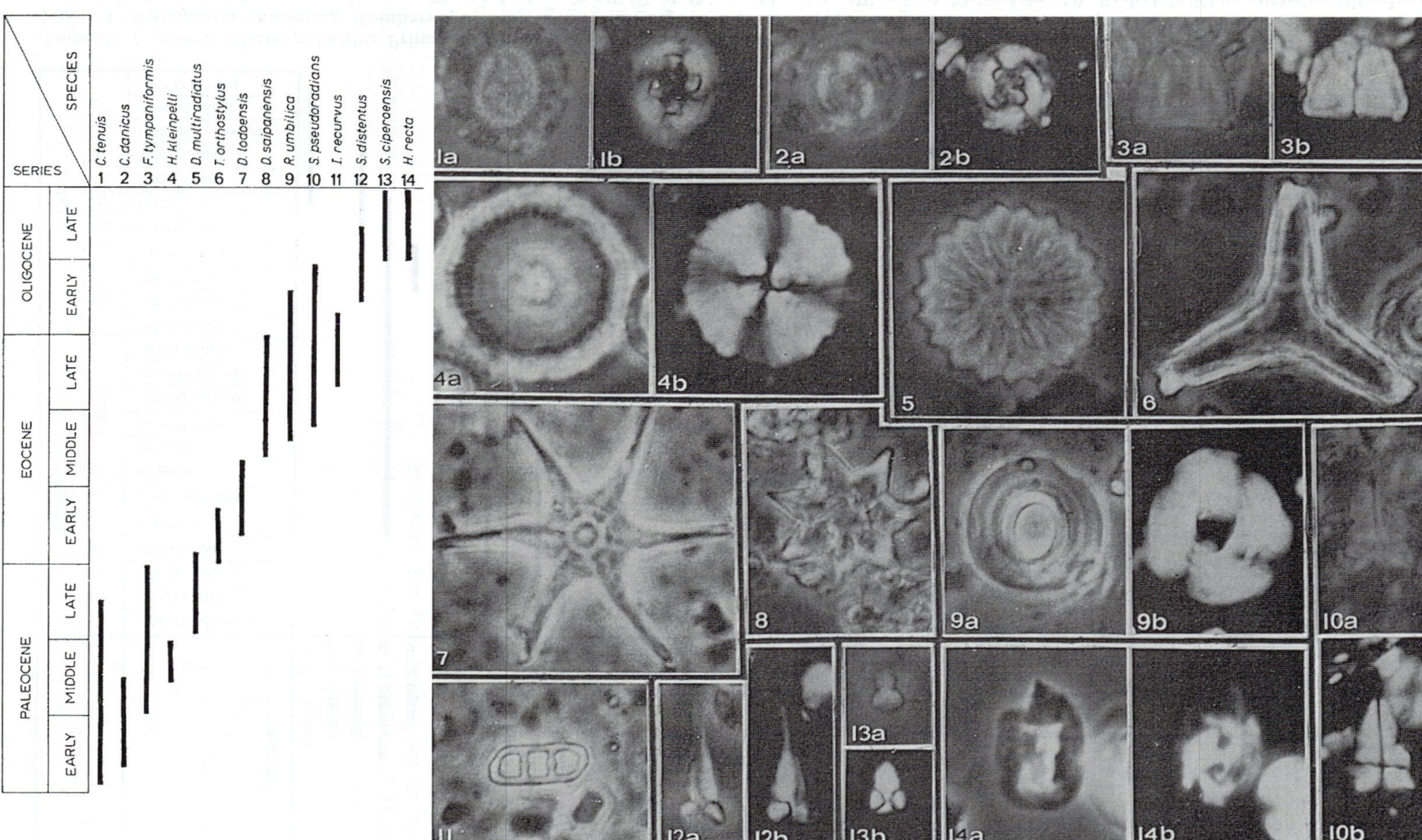

Legend: *1*, *Cruciplacolithus tenuis* (Stradner) Hay and Mohler; *2*, *Chiasmolithus danicus* (Brotzen) Hay and Mohler; *3*, *Fasciculithus tympaniformis* Hay and Mohler; *4*, *Heliolithus kleinpelli* Sullivan; *5*, *Discoaster multiradiatus* Bramlette and Riedel; *6*, *Tribrachiatus orthostylus* Shamrai; *7*, *Discoaster lodoensis* Bramlette and Riedel; *8*, *D. saipanensis* Bramlette and Riedel; *9*, *Reticulofenestra umbilica* (Levin) Martini and Ritzkowski; *10*, *Sphenolithus pseudoradians* Bramlette and Wilcoxon; *11*, *Isthmolithus recurvus* Deflandre; *12*, *Sphenolithus distentus* (Martini) Bramlette and Wilcoxon; *13*, *S. ciperoensis* Bramlette and Wilcoxon; *14*, *Helicosphaera recta* (Haq) Jafar and Martini.

TABLE IV

Stratigraphic ranges and light microscopic illustrations of selected late Cenozoic nannofossil species (figures *a* are under phase-contrast and *b* under cross-polarized light conditions; all figures × 3,200)

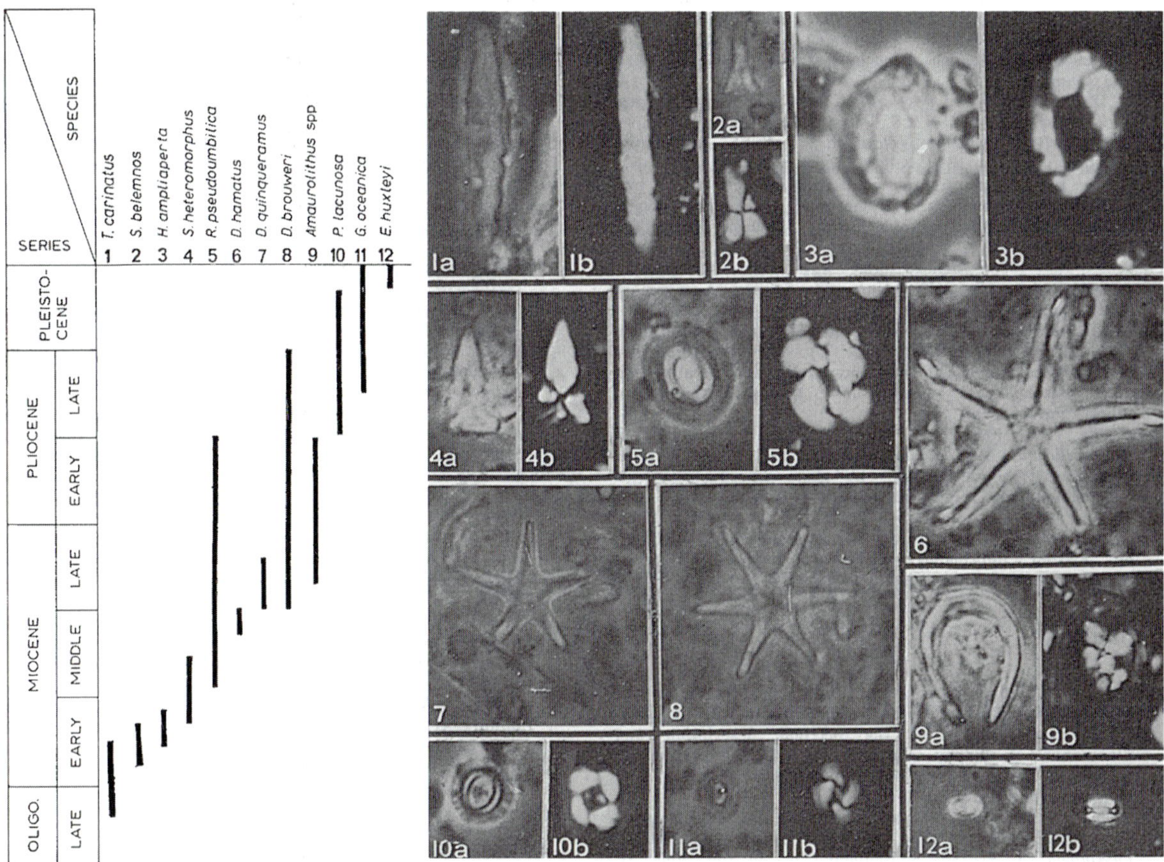

Legend: *1*, *Triquetrorhabdulus carinatus* Martini; *2*, *Sphenolithus belemnos* Bramlette and Wilcoxon; *3*, *Helicosphaera ampliaperta* Bramlette and Wilcoxon; *4*, *Sphenolithus heteromorphus* Deflandre; *5*, *Reticulofenestra pseudoumbilica* (Gartner) Gartner; *6*, *Discoaster hamatus* Martini and Bramlette; *7*, *D. quinqueramus* Gartner; *8*, *D. brouweri* Tan Sin Hok; *9*, *Amaurolithus delicatus* Gartner and Bukry (notice the almost complete extinction under cross-polarized light); *10*, *Pseudoemiliania lacunosa* (Kamptner) Gartner; *11*, *Gephyrocapsa oceanica* Kamptner; *12*, *Emiliania huxleyi* (Lohmann) Hay and Mohler.

open ocean. For a scheme more readily applicable to the oceanic sections zonations suggested by Bukry (1971 and 1973) should be consulted. For a Paleogene zonation of the relatively higher latitudes of New Zealand and adjacent Southern Ocean areas the reader is referred to Edwards (1971).

As compared to the Cenozoic rates, the nannoplankton evolutionary rates were relatively slower through most of the Mesozoic and thus the zonations for this era are not as refined. Prins (1969) suggested a zonation for the early Jurassic; Thierstein (1973) for the early Cretaceous andC epek and Hay (1969) and Manivit (1971) for the late Cretaceous. The reader is referred to Thierstein (1976) for a complete summary of the existing knowledge on the Mesozoic nannofossil zonation.

To make this text biostratigraphically more useful for the reader, range-charts of key species and their light microscopic illustrations are included in Tables II–IV. As far as possible those species with well-known ranges which are fairly common and can be more readily recognized have been chosen to present three- or four-fold subdivisions of each epoch (see Table II for the Mesozoic; Table III for the early Cenozoic and Table IV for the late Cenozoic).

Sexual dimorphism

There exists no other group of fossil invertebrates whose sexual dimorphism has been so thoroughly studied as the ostracodes. Dimorphism occurs not only in shape and function of the soft body parts, size, shape or sculpture of the carapace, but may manifest itself also in finer structural details, such as the pattern of muscle scars or even in behavior and habitat. The dimorphism of ostracode carapaces may be classified into two categories: (1) the domiciliar dimorphism which affects the size and shape of the carapace; and (2) the extradomiciliar (sculptural) dimorphism in which the proper carapace cavity (**domicilium**) is not affected. Dimorphic features are usually absent in larvae, although a weak sexual dimorphism is sometimes noticeable during the later larval stages.

Determination of the sex of fossil ostracodes is facilitated when dimorphic features can be compared with those of living taxa. A second criterion for sex recognition is the sex ratio. Among living ostracodes both sexes may be equally numerous, but more often the percentage of females is higher than that of males. Similar ratios may be observed in fossil ostracode populations.

Females may be larger, equal in size, or smaller than the males. The two sexes may also differ in lateral outline or length/height ratio (Fig. 23). Strong dimorphism is shown by species with brood-care. In modern ostracodes and their fossil relatives the brood pouches are situated in the posterior part of the carapace so that the female domicilium is characterized by an inflation. This type is called **kloedenellid**, also **cytherellid dimorphism**. In Paleozoic beyrichiomorphs the brood pouches termed **cruminae** (singular: **crumina**) are anterioventral or centroventral in position, and the type of dimorphism is called **cruminal dimorphism** (Fig. 24).

Different types of extradomiciliar dimorphism are abundant and strongly developed in Paleozoic hollinaceans (Fig. 25). An important role in this group is played by the two adventral structures, **velum** and **histium** (Fig. 16), both of which may be dimorphic. The part of the velum or histium which is modified in adult females when compared with adult males has been called the **dolon**. In females of the hollinacean ostracodes the dolo-

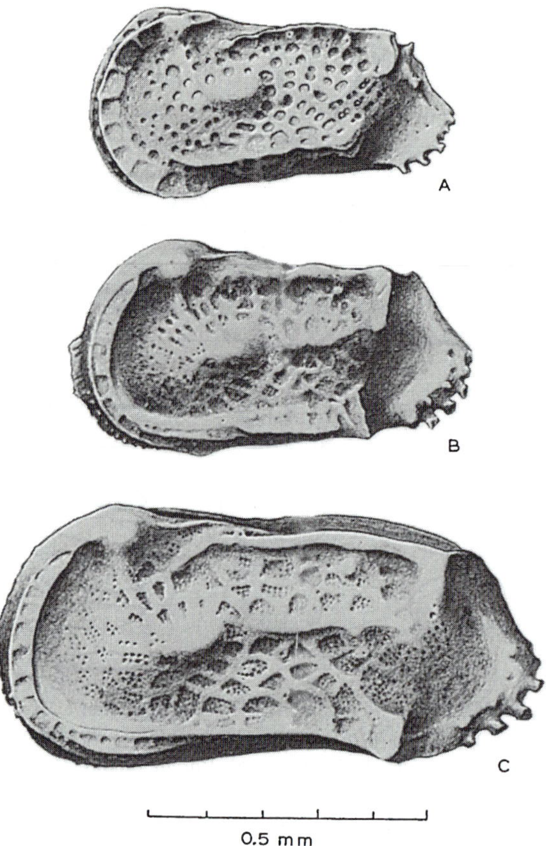

Fig. 23. Sexual dimorphism and difference in larval and adult sculpture in the trachyleberidid *Rehacythereis* (?) *kodymi* (Pokorný) from the Upper Cretaceous of Czechoslovakia. A, larval stage; B, female; C, male. (After Pokorný, 1967a.)

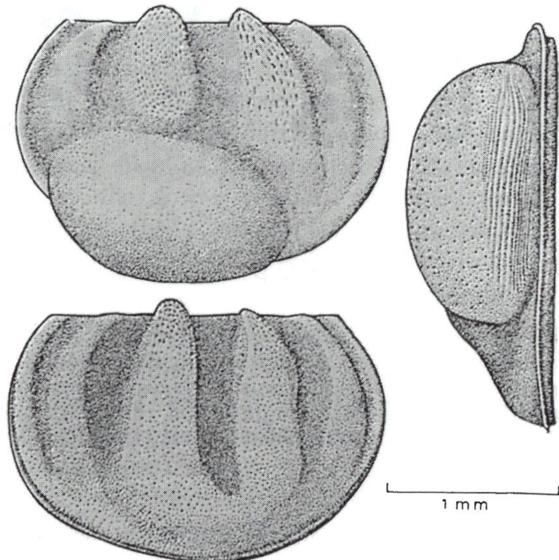

Fig. 24. Cruminal dimorphism in beyrichiomorphid ostracodes, shown by *Londinia reticulifera* Martinsson from the Silurian of Sweden. Above and right: heteromorph; below; tecnomorph. (After Martinsson, 1963.)

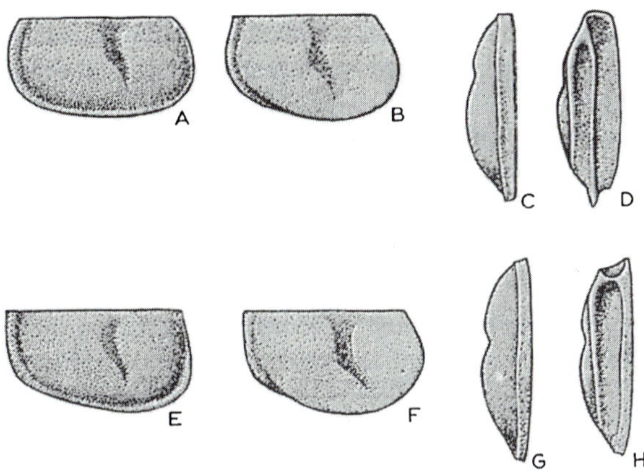

Fig. 25. Antral dimorphism in Ordovician hollinacean (tetradellid) genera. A–D, *Pentagona pentagona* (Jaanusson), a species with biantral dimorphism; A, C, lateral and ventral view of right tecnomorphic valve; B, D, lateral and ventral views of right heteromorphic valve. E–H, *Sigmobolbina variolaris* (Bonnema), a species with supravelar antral dimorphism; E,G, lateral and ventral views of a right tecnomorphic valve; F, H, lateral and ventral views of a right heteromorphic valve. Not to scale. (After Jaanusson, 1966.)

nal portions of adventral structures border concave areas termed **antra** (singular: **antrum**). When situated between the marginal ridge and the velum, the are called **infravelar antra**, when between the velum and histium, **supravelar antra**. Accordingly, the dimorphism is classified by Jaanusson (1966) as infravelar, supravelar or biantral (Fig. 25). The antrum may be simply channel-like, or may be partitioned into separate loculi. In modern ostracodes sculptural dimorphism is generally much less strong and expressed by dimorphic development of ridges, spines, etc.

Because of problems of sex recognition in the large group of predominantly Paleozoic beyrichiocopids, a neutral designation of the two dimorphs has been considered useful. The term **tecnomorph** has been introduced for carapaces of larval stages and those of adults which are essentially similar to larval stages. Those adult carapaces displaying features not present in tecnomorphs have been termed **heteromorphs** and are supposed to be females.

ECOLOGY

Ostracodes probably originated in a marine environment and the largest number of species still inhabits the pelagic and benthic realms of the ocean from the shoreline down to several thousand meters, and from the equator to polar seas. Some species flourish in brackish waters and some are found even in hypersaline environments. Ostracoda have also undergone ecologic radiation in fresh-water environments, from which they are known since the Carboniferous. Some lineages of both fresh-water and marine ostracodes have even invaded terrestrial niches, living in the moist humus of the forests (Harding, 1955), in the aerial part of the fresh-water floating plant accumulations (Danielopol and Vespremeanu, 1964), or, as reported by Schornikov (1969) for the aptly named genus *Terrestricythere*, in the fine, wet gravel on the sea shore of the Kuril Islands.

Nutrition

Ostracodes have evolved a wide variety of nutritional systems including filter-feeding and deposit-feeding. Numerous species feed on marine plants and small living animals such as annelids, turbellarians, nemerteans, or small crustaceans. Some eat detritus from decaying vegetal or animal tissues, while others are **limnivorous**, eating bottom sediments without any selection. Some have their oral apparatus transformed into piercing and sucking organs which are used for the intake of plant juices. Other species of the same family were observed to suck on dead polychaetes, amphipods and other animals. About thirty ostracode genera are known to be commensals, clinging to the appendages or gill cavities of other crustaceans and to the body surface of echinoderms. Some ostracodes interpreted as parasitic were described from the gills and nostrils of fishes (Harding, 1966). Others have glands along the valve margins which secrete a sticky substance to which food adheres. This is then brushed off by mandibular palps and brought to the mouth.

Distribution of marine ostracodes

A comparatively small number of marine ostracodes inhabit the pelagic realm, some living in surficial waters, others distributed through the water column. The greatest

(Fig. 48B) are usually small, mostly with a pronounced caudal process, often with ventrolateral extensions. Marine to oligohaline. Upper Triassic to Recent. The **leptocytherids** (Fig. 48C) include genera with low reniform carapaces, merodont hinges, mostly irregularly running line of concrescence, branched marginal pore cannals. Mostly marine to brackish waters. Tertiary to Recent. The **cytherideids** (Fig. 49A) include genera with strongly calcified carapaces, with broadly rounded anterior and usually narrowly rounded to acute posterior margins. Inner margin and line of concrescence mostly coincident, sometimes narrow vestibula developed. Marginal pore canals usually densely arranged, simple. Lateral pore canals of sieve-type. Marine to fresh water. Jurassic to Recent. The **loxoconchids** (Figs. 49B) have a characteristic rhomboid shape. Their marginal zone is wide, traversed by a few simple straight pore canals. Lateral pore canals are of sieve type. Marine to oligohaline. Jurassic to Recent. The **xestoleberidids** (Fig. 49C) have a subovate, swollen, smooth, or punctate carapace with arched dorsum. Their most characteristic feature is the "xestoleberidid spot" in the eye region. Most are shallow marine inhabitants. Cretaceous to Recent. The **krithids** (Fig. 49D) include species with elongate to highly arched carapaces. The calcified part of the inner lamella is broad, often with irregularly running inner margin. A pocket-like vestibulum is often developed at the anterior margin. Marine, shallow water to hadal. Upper Cretaceous to Recent. The **progonocytherids** (Fig. 50A, B), as originally defined, comprised genera having an archidont hinge, i.e. quadripartite hinge in which the anteromedian part is more coarsely crenulate than the posteromedian. Radial pore canals are straight, not abundant. Triassic to Cretaceous. The **protocytherids** (Figs. 50C, D) have pear-shaped to quadrangular, usually, strongly calcified carapaces, and broad

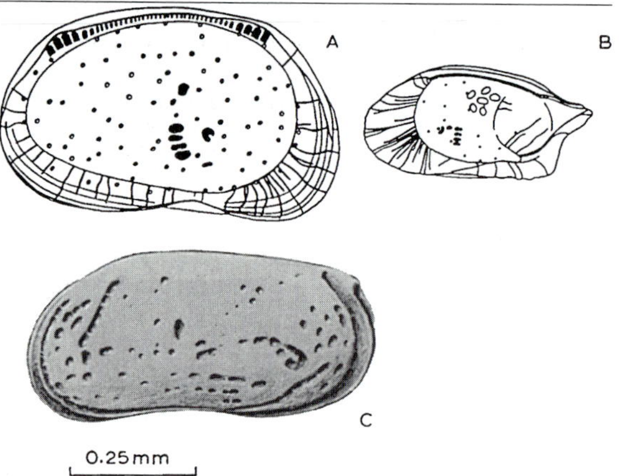

Fig. 48. A. *Cythere lutea* (O.F. Müller), a cytherid. Scheme of a left female valve seen from the inside. Holocene, The Netherlands. × 60. (After Wagner, 1957.) B. *Semicytherura angulata* (Brady), a cytherurid. Scheme of the right valve from the inside. Holocene, The Netherlands. × 60. (After Wagner, 1957.) C. *Leptocythere pellucida* (Baird), a leptocytherid. Carapace from the left side. Recent, England.

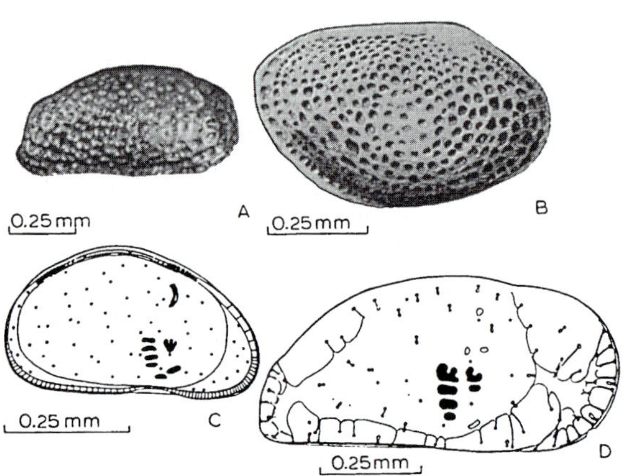

Fig. 49. A. *Cytheridea acuminata* Bosquet, a cytherideid. Right valve, Miocene, Czechoslovakia. B. *Loxoconcha bairdii* G.W. Müller, a loxoconchid. Right valve, Recent, Gulf of Naples, Italy. C. *Xestoleberis aurantia* (Baird), a xestoleberidid. Scheme of the left valve seen from the interior. Note the "xestoleberidid spot" in the anterodorsal region. Holocene, The Netherlands. (After Wagner, 1957.) D. *Krithe undecimradiata* Ruggieri. Male, right valve. Quaternary, Italy. (After Greco and others, 1974.)

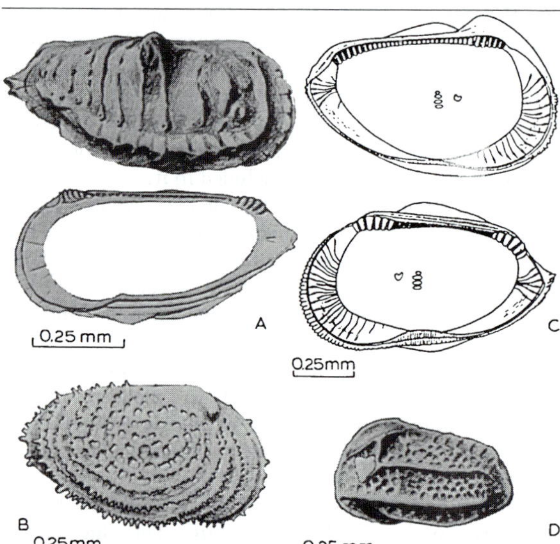

Fig. 50. A and B, progonocytherids. A. *Lophocythere propinqua* Malz. Right valve from the outside and from the inside. Middle Jurassic, France. B. *Centrocythere denticulata* Mertens. Right valve. Lower Cretaceous, England. C and D, protocytherids. C. *Protocythere triplicata* (Roemer). Left and right female valves from the inside. Lower Cretaceous, German Federal Republic. (After Triebel, 1938.) D. *Pleurocythere impar* Triebel. Left valve, Middle Jurassic.

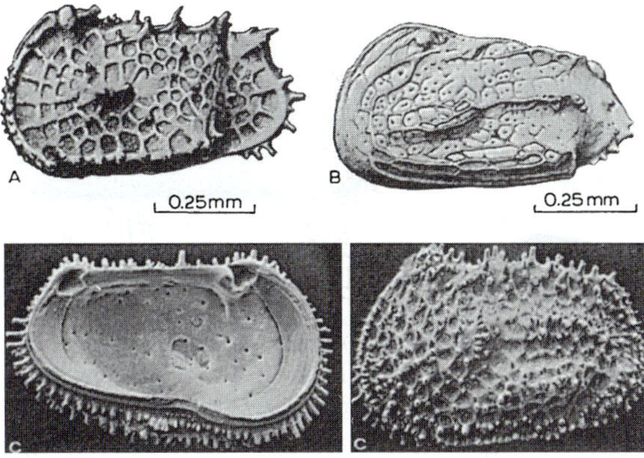

Fig. 51. The trachyleberidids. A. *Oertliella reticulata* (Kafka). Female, carapace from the left side. Upper Cretaceous, Bohemia, Czechoslovakia. (After Pokorný, 1964a.) B. *Mosaeleberis interruptoidea* (Van Veen). Left valve. Upper Cretaceous, Bohemia, Czechoslovakia. C. *Henryhowella asperrima* (Reuss). Pliocene, Italy. Left: interior; right: exterior of left valve. SEM photos. (Courtesy R.H. Benson.)

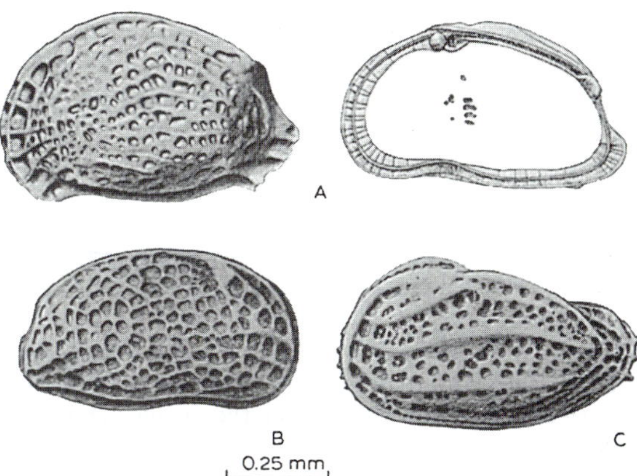

Fig. 52. A and B, hemicytherids. A. *Pokornyella limbata* (Bosquet). Left: left valve from the outside; right: interior of right valve. Paleogene, Paris Basin, France. × 45. (After Keij, 1957.) B. *Hemicythere villosa* (Sars). Right valve, Skagerrak, Sweden. C. *Cytheretta gracillicosta* (Reuss), a cytherettid. Left valve; Paleogene, German Democratic Republic.

zone of concrescence traversed by long and sinuous simple marginal pore canals. Sculpture is basically with three longitudinal ribs. Jurassic to Recent, with many good index species in the Middle Jurassic through Lower Cretaceous. The **trachyleberidids** (Fig. 51) have mostly strongly calcified carapaces, frequently strongly ornate, amphidont hinges, usually a V-shaped frontal scar, a well-developed zone of concrescence with fairly numerous to abundant marginal pore canals. Marine, shallow water to abyssal. Middle Jurassic to Recent. The trachyleberidids are among those families which have the best fossil record and are most important for the stratigraphy and paleoecology of the Cretaceous and the Cenozoic. The **hemicytherids** (Fig. 52A, B) undoubtedly arose from the trachyleberidids. Instead of a V-shaped frontal muscle scar characteristic for most trachyleberidids, all hemicytherids have two to three frontal scars and some of their adductor scars are often secondarily divided. The oldest genera appeared towards the end of the Cretaceous, but the family belongs to the most characteristic taxa of the shallow-water Cenozoic assemblages. Mostly marine, not uncommon in brackish waters, exceptionally in fresh water. The **cytherettids** (Fig. 52C) are phylogenetically close to the trachyleberidid-hemicytherid group. Their carapaces are oval to cylindrical, their most characteristic feature is the extraordinarily broad zone of concrescence. Inner margin coincident with the line of concrescence. Hinge amphidont, lateral pore canals simple. Upper Cretaceous to Recent.

(ii) The **bairdiaceans** (Fig. 53) have changed very little since early Paleozoic times and are a good example of what is called the "living fossils". A calcified inner lamella is present even in oldest forms. From their central longeval lineage short-lived offshoots were repeatedly derived. The superfamily attained its maximum development during the late Paleozoic and the Triassic. A small adaptive radiation occurred in the Cenozoic. They comprise one living and several fossil families:

The bairdiids (Fig. 53A, B) include covex-backed ostracodes with typical "bairdioid" shape. Their carapaces are smooth to highly ornate. Hingement short, of ridge-and-groove type. Ordovician to Recent. The **beecherellids** (Fig. 53C) include fairly elongate smooth ostracodes with straight to slightly vaulted

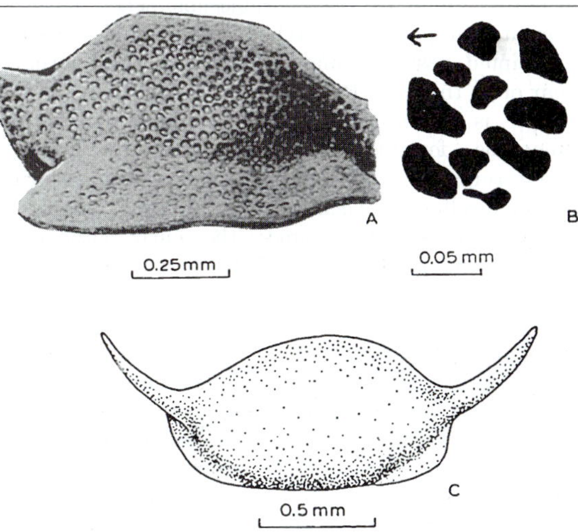

Fig. 53. A. *Havanardia havanensis* Pokorný, a bairdiid. Right valve. B. Adductor muscle scars of the same. Recent, off Havana, Cuba. (After Pokorný. 1968b.) C. *Acanthoscapha volki* Blumenstengel, a beecherellid. Left valve. Upper Devonian, German Democratic Republic. (After Blumenstengel, 1965.)

several characters: a long straight dorsal margin, a slightly convex ventral margin, an adductor scar pit ("kirkbyan pit") and reticulate surface. Ordovician to Triassic.

The position of the Neogene to Recent **punciids** (Fig. 59) is still under discussion. Some authors see their affinities to the hollinomorphs (eurychilinaceans), others to the kirkbyaceans.

IMPORTANCE OF OSTRACODES FOR PALEOGEOGRAPHY

Marine ostracodes are not as well suited for interregional and intercontinental stratigraphic correlation as are other groups of planktonic microfossils. Since benthic ostracode species have no planktonic larvae, the shallow, warm-water species cannot easily cross geographic barriers. This limitation, however, qualifies them as excellent paleobiogeographical markers. Ostracodes are also efficient tools for the study of paleobathymetry and paleosalinity. Hence, their study may be extremely helpful for tracing of paleogeographic changes.

Of considerable paleogeographical interest is the history of the cosmopolitan deep-sea fauna in the Mediterranean Province studied by Benson and Sylvester-Bradley (1971). Its characteristic elements are found in Paleocene to Middle Miocene and in Pliocene sediments from different areas of the Mediterranean province. In the late Miocene, the evolution of the normal marine ostracode fauna of the Mediterranean Sea was interrupted, as this sea was cut off from the Atlantic and transformed into a series of lagoons. Some of these dried up, others desalinified and developed a peculiar endemic fauna. This fauna had many elements in common with assemblages from the contemporary Paratethyan basins (see Fig. 26). The Paratethyan basins of late Miocene and Pliocene time were characterized by low salinities and endemic ostracode communities. These endemics are excellent indicators of changing communication between these basins, of their paleosalinities and are most useful as stratigraphic markers in the search for oil and coal within these basins.

At the beginning of the Pliocene, communication between the Mediterranean and Atlantic was re-established in the west, so that Atlantic euhaline species, even the deep-sea psychrospheric fauna, re-invaded the Mediterranean. As the connection between the Mediterranean Province and the Indo-Pacific region has been interrupted since the middle Miocene, Recent Mediterranean ostracodes are chiefly of Atlantic origin and differ substantially from their Tertiary forerunners. Towards the end of the Tertiary the shallowing of the sill at the Straits of Gibraltar prevented the entrance of cold deep oceanic waters into the Mediterranean, leading to the elimination of the psychrosphere and consequently of the psychrospheric fauna from this area. The present Mediterranean fauna is entirely thermospheric as the temperature even at abyssal depths is near 13°C.

The present Mediterranean has another aspect of great interest to a biogeographer. With the opening of the Suez Canal more than a hundred years ago, and especially through its reconstruction in the years after World War II, a seaway has been established between two large provinces: the Indo-West Pacific and the Mediterranean. The main stream of immigrants goes from the Red Sea to the Mediterranean and includes shallow-water ostracodes. McKenzie (1973) pointed out that in older collections from the Mediterranean some elements are absent which were observed in recent collections and are thus probably of Red Sea origin. This is a special case of increasing human affect on the composition of naturally established regional faunas, as our seaways and extensive sea traffic tend to introduce many new elements into what were indigenous faunas.

In the Neogene of the Caribbean region Van den Bold (1974) was able to distinguish two faunal provinces, and three subprovinces. Some Caribbean Cenozoic shallow-water ostracodes even reached the remote Galapagos Islands, probably by dispersal on drifting objects, in a warm-water current system. Two of these lineages underwent a spectacular insular evolutionary radiation on the Galapagos (Fig. 60).

A similar colonization by long distance, rather chance dispersal, called **sweepstakes** routes, has been demonstrated also for the islands of the equatorial Pacific. According to McKenzie (1969), several genera disappear progressively as we go east in the Pacific from New Caledonia to Fiji to Samoa and finally to

Fig. 60. Insular radiation of the hemicytherid genus *Radimella* Pokorný in the Galapagos Islands. The individuals of this shallow-water genus display a discontinuous variation of sculpture which speaks of the existence of genetically discrete populations. A larger number of these discrete forms occur at the same locality, suggesting their genetic isolation. The distinction of such closely similar forms is possible only by careful morphologic and biometric analysis. (After Pokorný, 1970.)

Bora Bora near Tahiti. Similarly the endemism of the New Zealand fauna has been affected by its position at the end of a sweepstakes route.

Information about paleocirculation patterns has been derived from the study of ostracodes. The importance of the West-Wind drift for the distribution of Austral-Asian ostracodes has been underlined by McKenzie (1973). Because of continental drift, this current was not operative until late Cretaceous time, and in some parts of the area not until the Oligocene. According to McKenzie, the South Equatorial Current which flows from West Africa to the Caribbean via the coasts of Brazil and Venezuela has acted as another sweepstakes route and is responsible for some elements common to the East African and Caribbean provinces.

Ostracodes have been used also as paleoclimatic indicators. During the Quaternary some north European Atlantic species temporarily penetrated the Mediterranean, so that the occurrence of these "northern guests" has been used for the recognition of cold climatic periods (Ruggieri, 1971).

Hazel (1970, 1971) made detailed studies of Recent bottom samples in the region from Nova Scotia to Long Island where many sublittoral, cryophilic ostracode species are not

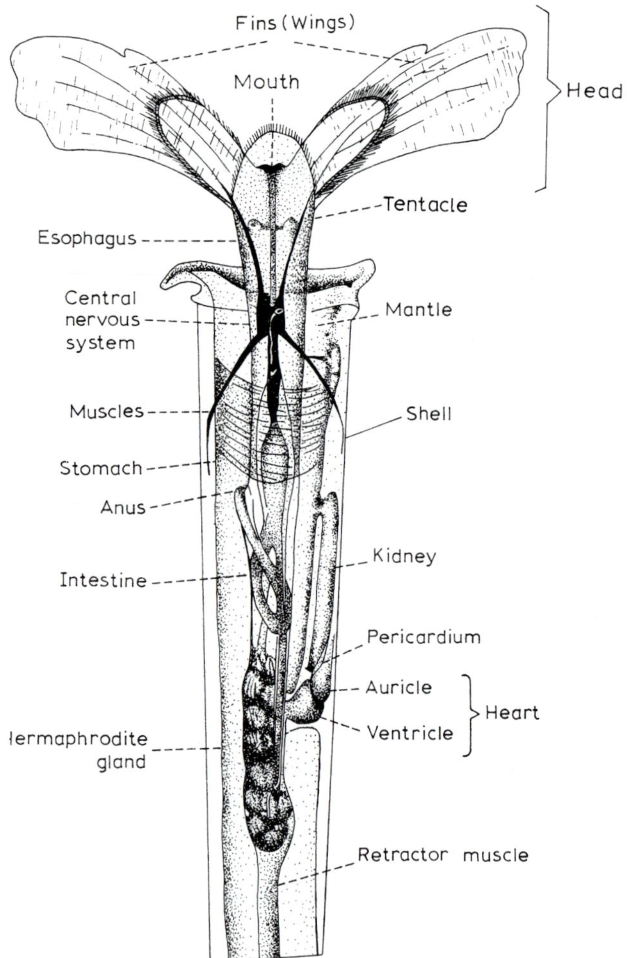

Fig. 1. Morphology of soft parts in *Creseis* (after Tregouboff and Rose, 1957)

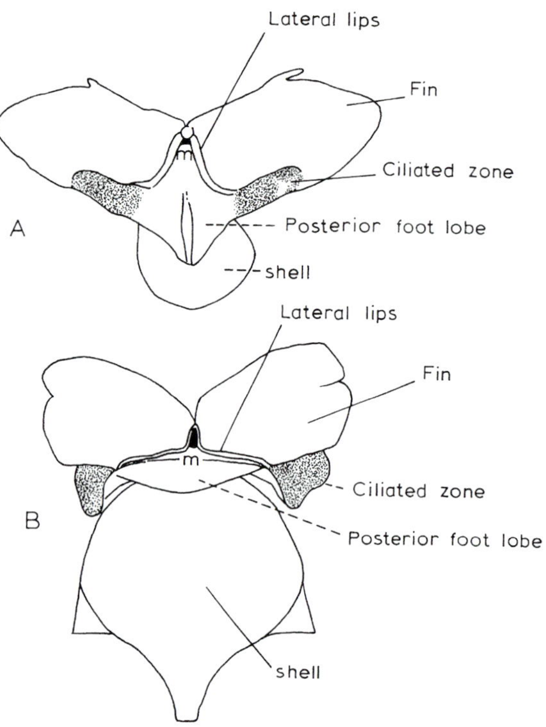

Fig. 2. Diagrammatical view of the organization of Euthecosomata. Ventral view. A. Limacinidae. B. Cavoliniidae. (After Meisenheimer, 1905.) *m* = mouth.

The reproductive system. All the Opistobranchia are hermaphroditic, meaning that each individual has both male and female sexual organs and is capable of producing both ova and spermatozoa. A period of male maturity during which spermatozoa are generated, precedes that of female maturity. Both ova and spermatozoa are formed in one reproductive organ, the **hermaphrodite gland**.

The nervous system. This consists of a number of nerve centers or ganglia, joined by nerve cords and communicating by means of nerves with various parts of the organism. The cerebral ganglia supply nerves to the head and tentacles; the pedal ganglia supply the foot lobes and fins; and the visceral ganglia supply nerves to the mantle, heart, kidney, and the gonads.

The circulatory system. The heart consists of an auricle and a ventricle, enclosed in the pericardial sac. Blood is pumped from the heart through numerous vessels to various parts of the body. The kidney is the main organ for excretion of metabolic waste products; it adjoins both the digestive gland and the pericardium and opens into the mantle cavity.

The mantle. The mantle is an overgrowing sheet of tissue. It appears that special structures in the mantle edge may be responsible for the shell formation or secretion. The space between the mantle and the underlying tissues, known as the mantle cavity, constitutes the respiratory chamber; gills are present in only one genus.

ECOLOGY

Pteropods are exclusively marine and generally live in the open ocean, swimming in the uppermost 500 m; however, some forms are known to live at great depths.

The present-day distributional patterns of pteropods are fairly well known. They are ubiquitous and abundant and about eighty species and subspecies inhabit the world's oceans. Their distribution is controlled by various physical and chemical parameters of the environment, such as temperature, salinity, food, oxygen and water depth.

Temperature

Temperature is the main factor governing the distribution of pteropods. At present, well-defined latitudinal temperature gradients exist; from the cold polar regions, temperatures increase progressively toward the equator. This gradual water temperature change is reflected in pteropod population composition. Thus, the two polar seas are inhabited by one pteropod species, *Limacina helicina* Phipps, whereas the warm, tropical zones are populated by many genera and species (Table I).

Salinity

Marine holoplanktonic invertebrates are cold-blooded and have body fluids isotonic with the surrounding water. For this reason they are limited to the narrow salinity ranges of oceanic water. The average sea-water salinity varies between 35‰ and 36‰. As mentioned in the previous paragraph, the warm regions of the three major oceans, Atlantic, Indian and Pacific, support a diversified pteropodal fauna. However, in land-locked warm seas where evaporation exceeds precipitation and runoff from land, salinities are much higher than in the open ocean. In the Red Sea with surface-water salinities greater than 40‰ but with temperatures similar to those of the oceans, only about 50% of the oceanic species are known to occur (Herman-Rosenberg, 1965). In the Mediterranean where salinities are intermediate between the Red Sea and the open ocean and temperatures are similar to the oceans at comparable latitudes, about 75% of the open-ocean species

TABLE I

Present-day distribution of some pteropods in the oceans

Warm	Cold-temperate	Cold-polar
Cavolinia gibbosa		
C. globulosa		
C. inflexa		
C. longirostris		
C. tridentata		
C. uncinata		
Clio cuspidata		
C. polita		
C. pyramidata convexa	*C. pyramidata pyramidata*	
Creseis acicula		
C. virgula		
C. conica		
Cuvierina columnella		
Diacria quadridentata		
D. trispinosa		
Hyalocylix striata		
Styliola subula		
Limacina bulimoides		
L. inflata		*Limacina helicina*
L. lesuerii		
L. trochiformis	*Limacina retroversa*	
Peraclis spp.		

TABLE I

Correlation between calpionellid, ammonite, and calcareous nannoplankton zones

Periods/stages			Ammonite zones	Calpionellid zones	W.A.	Nannoplankton zones
LOWER CRETACEOUS	Valanginian	Upper				*Calcicalathina oblongata*
		Upper	*verrucosum*	- - - - - - -	- - -	*Calcicalathina oblongata*
		Lower	*campylotoxus*	*Calpionellites*	E	*Calcicalathina oblongata*
		Lower	*roubaudi*	*Calpionellites*	E	*Calcicalathina oblongata*
		Lower	*pertransiens*	*Calpionellites* / *Calpionellopsis*	E / D	*Calcicalathina oblongata* / *Cretarhabdus crenulatus*
	Berriasian		*boissieri*	*Calpionellopsis*	D	*Cretarhabdus crenulatus*
	Berriasian		*occitanica*	*Calpionella*	C	*Nannoconus colomi*
	Berriasian		*grandis*	*Calpionella*	B	*Nannoconus colomi*
UPPER JURASSIC	Tithonian	Upper	*jacobi*	*Calpionella*	B	?
		Upper	"*Durangites*"	*Crassicollaria*	A	*Conusphaera mexicana*
		Upper	*microcantha*	*Crassicollaria*	A	*Conusphaera mexicana*
		Lower/Middle	*ponti*	*Chitinoidella*	Chit.	*Conusphaera mexicana*
		Lower/Middle	*fallauxi*	?	?	*Conusphaera mexicana*

W.A.: calpionellid zones of the Western Alps, after Remane. Ammonite zones are at present mostly tentative: after Thieuloy (1974) for Valanginian; Le Hégarat (1971) for Berriasian; and Enay and Geyssant (1975) for Tithonian. Nannoplankton zones after Thierstein (1975). For faunal associations of calpionellid zones see Fig. 8.

Fig. 10. Important calpionellid species, in this section. ca. × 420.
A. *Chitinoidella boneti* Doben. Basal to middle Upper Tithonian. B. *Praetintinnopsella andrusovi* Borza. Wall structure predominantly hyaline, whereas the collar is still entirely microgranular. Basal Upper Tithonian. C. *Crassicollaria intermedia* (Durand Delga). Lower part of Upper Tithonian. D. *Crassicollaria brevis* Remane. Lower part of Upper Tithonian. E. *Crassicollaria massutiniana* (Colom). Lower part of Upper Tithonian. F. *Crassicollaria parvula* Remane. Upper Tithonian–Lower Berriasian. G. *Calpionella alpina* Lorenz. Upper Tithonian–Berriasian. Large from typical for the lower part of the Upper Tithonian. H. *Calpionella alpina* Lorenz. Upper Tithonian–Berriasian. Smaller, spherical variety typical for the Jurassic–Cretaceous boundary beds. I. *Calpionella elliptica* Cadisch. Lower Berriasian. J. *Tintinnopsella carpathica* (Murgeanu and Filipescu). Upper Tithonian–Lower Valanginian. Large form, typical of Cretaceous (exceptional section showing the caudal appendage along its whole length). K. *Tintinnopsella carpathica* (Murgeanu and Filipescu). Upper Tithonian–Lower Valanginian. L. *Calpionellopsis simplex* (Colom). Middle–Upper Berriasian. M. *Calpionellopsis simplex* (Colom). Middle–Upper Berriasian. N. *Calpionellopsis oblonga* (Cadisch). Upper Berriasian. Axial section with badly preserved collar. O. *Calpionellopsis oblonga* (Cadisch). Upper Berriasian. Oblique section with well-preserved collar. P. *Calpionellites darderi* (Colom). Lower Valanginian.

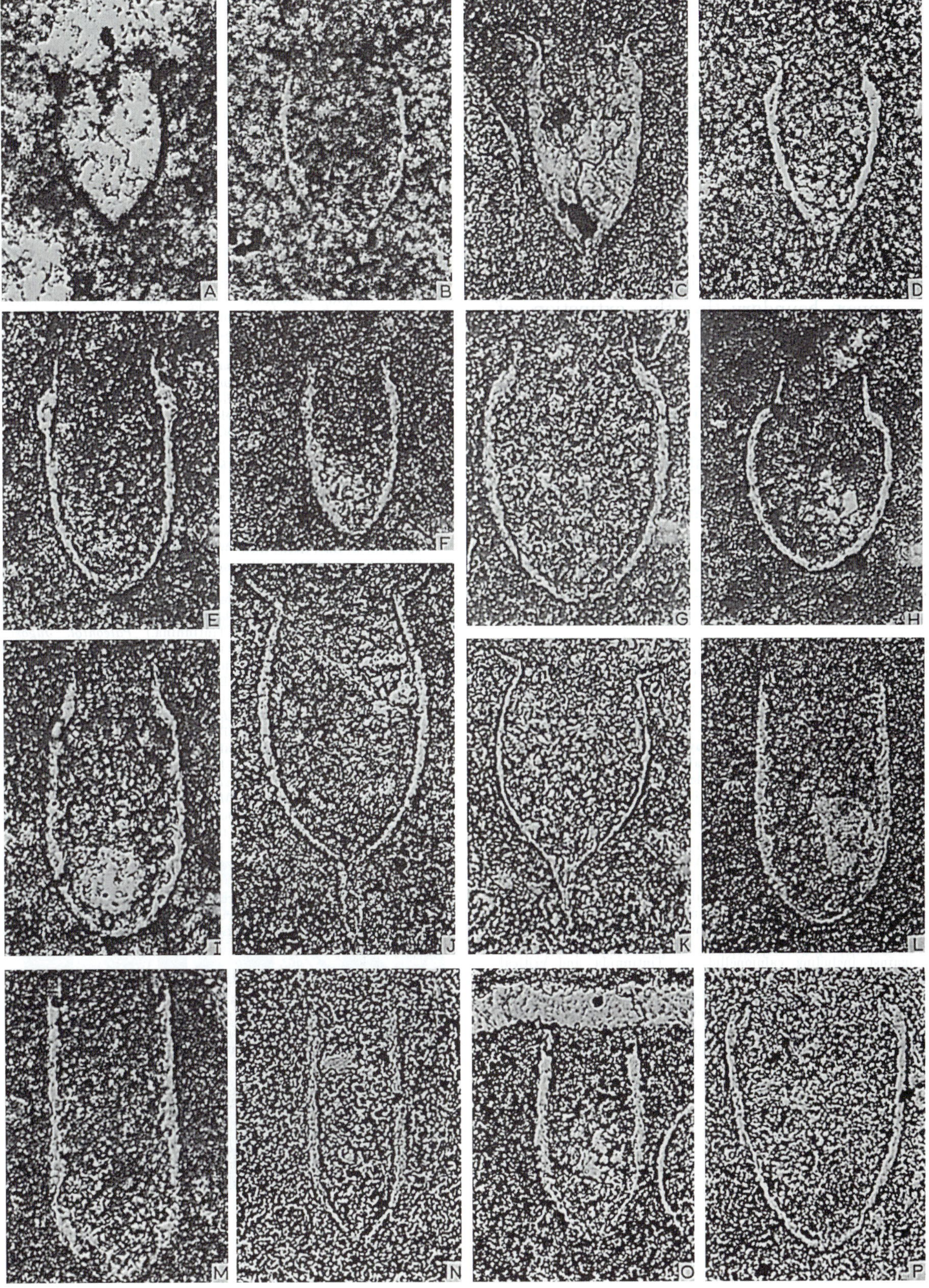
A
B
C
D
E
F
G
H
I
J
K
L
M
N
O
P

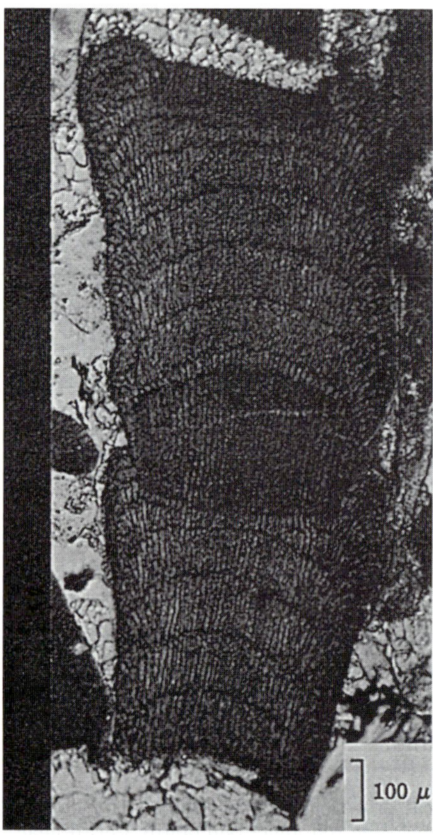

Fig. 9. Articulated-segmented coralline alga *Corallina* showing arcuate rows of cellular tissue (hypothallium) within segments. Miocene, Saipan.

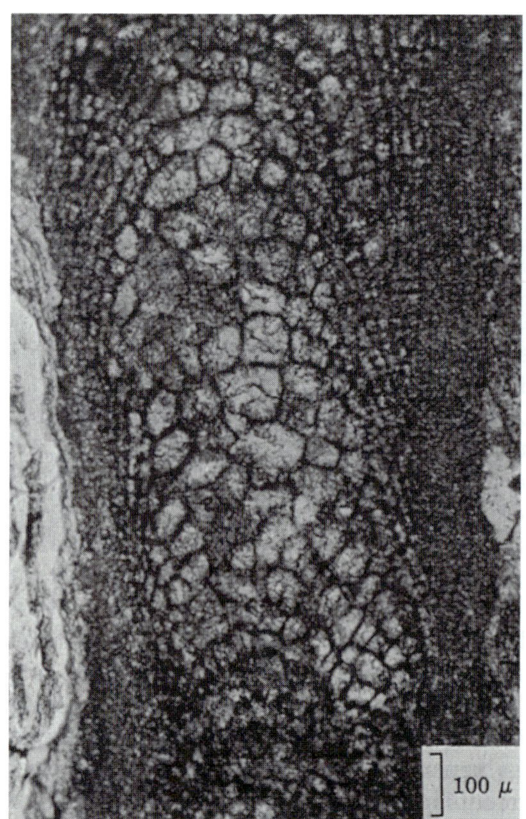

Fig. 11. Ancestral coralline alga *Archaeolithophyllum* showing arrangement of cells in hypothallium (central portion) and outer perithallium. Pennsylvanian, Missouri.

Fig. 10. Solenoporacean red alga *Parachaetetes*. Vertical section of nodular growth form showing cellular tissue. Upper Devonian, Western Australia.

encrusting and erect growth habits, and are volumetrically important contributors to limestone development. *Archaeolithophyllum* (Fig. 11), ranging from late Mississippian to middle Permian, is one of the more common and distinctive taxa belonging to this extinct group. This crustose alga, possessing **conceptacles** (cavity-containing sporangia) and differentiated cellular tissue, is remarkably similar to the modern coralline genus *Lithophyllum*.

Green algae (Chlorophyta)

Carbonate-secreting habits have been developed by two families of marine green algae, the Codiaceae and the Dasycladaceae. Skeletal carbonates in the green algae are exceedingly variable; they include rather thoroughly calcified **thalli**, fragile skeletal elements composed of aragonite needles, and surficial deposits of calcium carbonate representing molds of the thallus.

The microfossils of both codiaceans and dasycladaceans consist mainly of whole or broken segments, rather than entire plants. The shape of individual segments, in addition to the internal organization of filaments and pores, are the bases for classifying individual taxa. Erect, stalked codiaceans with bushy crowns, such as *Penicillus*, frequently disaggregate completely into micron-sized aragonite crystals which cannot be identified taxonomically. This is of interest sedimentologically because the process is considered to be the origin of much of the carbonate mud and silt accumulating in Recent tropical and subtropical shallow marine environments (Stockman and others, 1967). Although the amount of calcium carbonate produced per plant is small, they are so numerous and grow so rapidly that they produce volumetrically significant amounts of sediment. Similar kinds of calcareous green algae were probably the source of fine-grained sediment in ancient shallow-shelf carbonate environments. Both codiaceans and dasycladaceans have a long geologic record, extending from the Cambrian.

Codiaceae

Fossil calcareous Codiaceae exhibit two distinct growth habits: (1) crustose or nodular forms; and (2) erect plants, commonly consisting of segmented branches (Fig. 12). Nodular forms of the Codiaceae occur most often in the Paleozoic and genera are distinguished on the basis of the character of internal branching tubular filaments. *Ortonella* and *Garwoodia* are two common Paleozoic genera.

Most Mesozoic and Cenozoic calcareous codiaceans are erect plants (Fig. 13) several centimeters high, generally segmented and branching, and possess an internal structure composed of interwoven tubular filaments (Fig. 14). The shape and segmentation of the overall plant and the internal organization and branching of the filaments differentiate individual taxa. The extant genus *Halimeda*, which appeared first in the Cretaceous, is an important representative of the erect calcareous codiaceans. Skeletal carbonate in

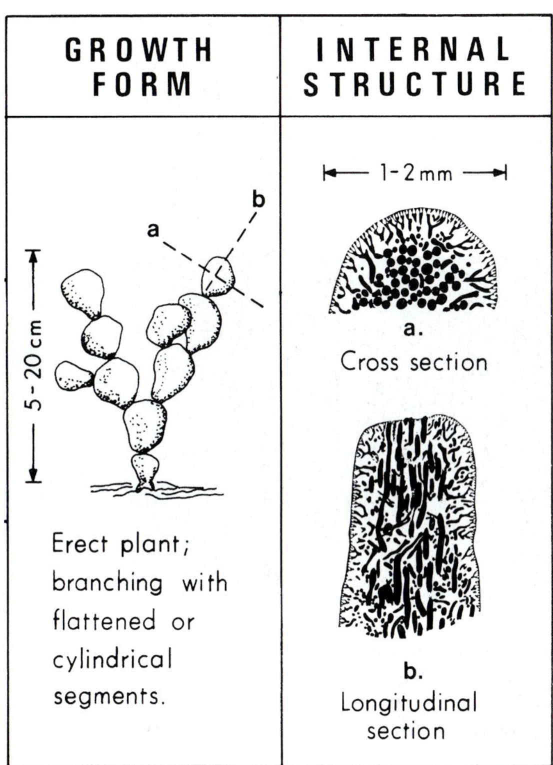

Fig. 12. Typical growth form and internal structure of erect calcareous Codiaceae (green algae).

Halimeda and related genera consists of minute elongate aragonite crystals. Calcification is more complete in older portions of the plant compared to younger parts, and outer regions of the thallus are more thoroughly calcified than the inner portion. These differences in degree of calcification are observed in both living and fossil specimens.

Dasycladaceae

Calcareous Dasycladaceae are erect, segmented, branching plants usually several centimeters high. Most species are characterized by a large central stem surrounded by tufts or whorls of smaller radiating branches; thus, dasycladacean skeletal remains are mainly molds of segments preserving arrangements of these features. Most fossils appear as hollow perforated cylinders or spheres (Figs. 15, 16, 17), while others are perforated discs or blade-like objects (Fig. 18). **Sporangia** (reproductive organs) are developed adjacent to the stem or branches and their outline may be preserved by calcification.

This group of calcareous green algae is

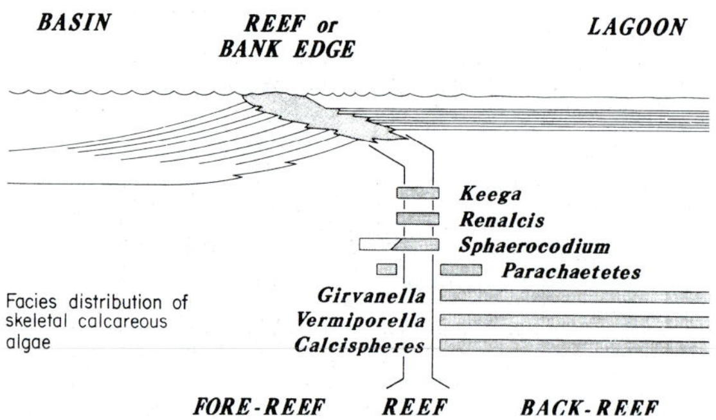

Fig. 25. Environmental distribution of principal taxa of skeletal calcareous algae in Upper Devonian reef complexes. *Keega* is a problematical ancestral coralline alga; *Parachaetetes* is a solenoporacean; *Renalcis*, *Sphaerocodium* and *Girvanella* are blue-green algae; *Vermiporella* is a dasycladacean; and calcispheres are believed to be reproductive bodies (gametangia) of unknown dasycladaceans. (From Wray, 1972.)

eous algae in Devonian carbonate reef complexes (Wray, 1972) indicate that individual taxa are restricted to particular depositional facies (Fig. 25). In this example, all of the algae are extinct and the relationship of most taxa to living groups in unclear; consequently, their paleoecology cannot be inferred from analogous living groups. Yet this empirical distribution pattern of Devonian calcareous algae within a facies complex does yield a comprehensible paleoenvironmental picture. Thus, calcareous algae can provide an important complement to other benthic organisms in determining biofacies in ancient carbonate shelf environments, despite the fact that the environmental limits of many fossil forms, at least at the lower taxonomic levels, have not been determined.

GEOLOGIC DISTRIBUTION

Major groups of benthic calcareous algae and many minor taxa have long geologic ranges. As a result, calcareous algae have limited value for age determinations and in biostratigraphy, although a few forms provide useful marker horizons in some parts of the section. Some algae, notably the primitive blue-greens, are characterized by extreme evolutionary conservatism, and living species are almost indistinguishable from species that lived millions of years ago. Also, in contrast to planktonic organisms, benthic calcareous algae were often restricted to narrow paleoecologic niches, and are not useful for widespread biostratigraphic correlations. Wray (1971) outlined the geologic distribution of calcareous algae considered to be important in ancient and modern reef development.

Stromatolites range from the Archaean; the oldest ones known (2.6 billion years) occur in southern Rhodesia. The late Precambrian of the Soviet Union has been subdivided into four zones based on stromatolite assemblages, and this chronology has been applied in other regions (Cloud and Semikhatov, 1969); however, many discrepancies in ranges and methodology have been noted, and a worldwide Proterozoic stromatolite zonation is open to criticism. Stromatolites of various ages have been used in the physical correlation of beds within sedimentary basins. The apparent decline in abundance of stromatolites in the Phanerozoic has been explained by the expansion of grazing and burrowing animals that destroyed algal laminae (Garrett, 1970).

The time distribution, relative abundance, and suggested evolution of the principal groups of marine skeletal calcareous algae are illustrated in Fig. 26. The time ranges of important genera are summarized in Fig. 27. From these one can recognize major trends in the occurrence of calcareous algae in time, and distinguish assemblages characteristic of particular time intervals.

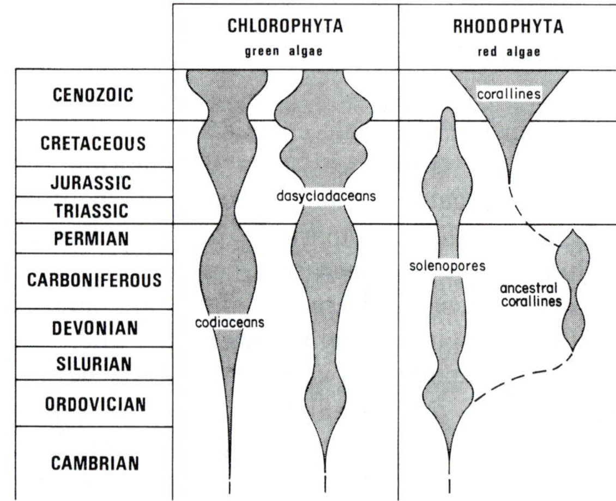

Fig. 26. Geologic distribution and inferred evolution of major groups of marine skeletal calcareous algae. Abundance and diversity is suggested by width of patterns.

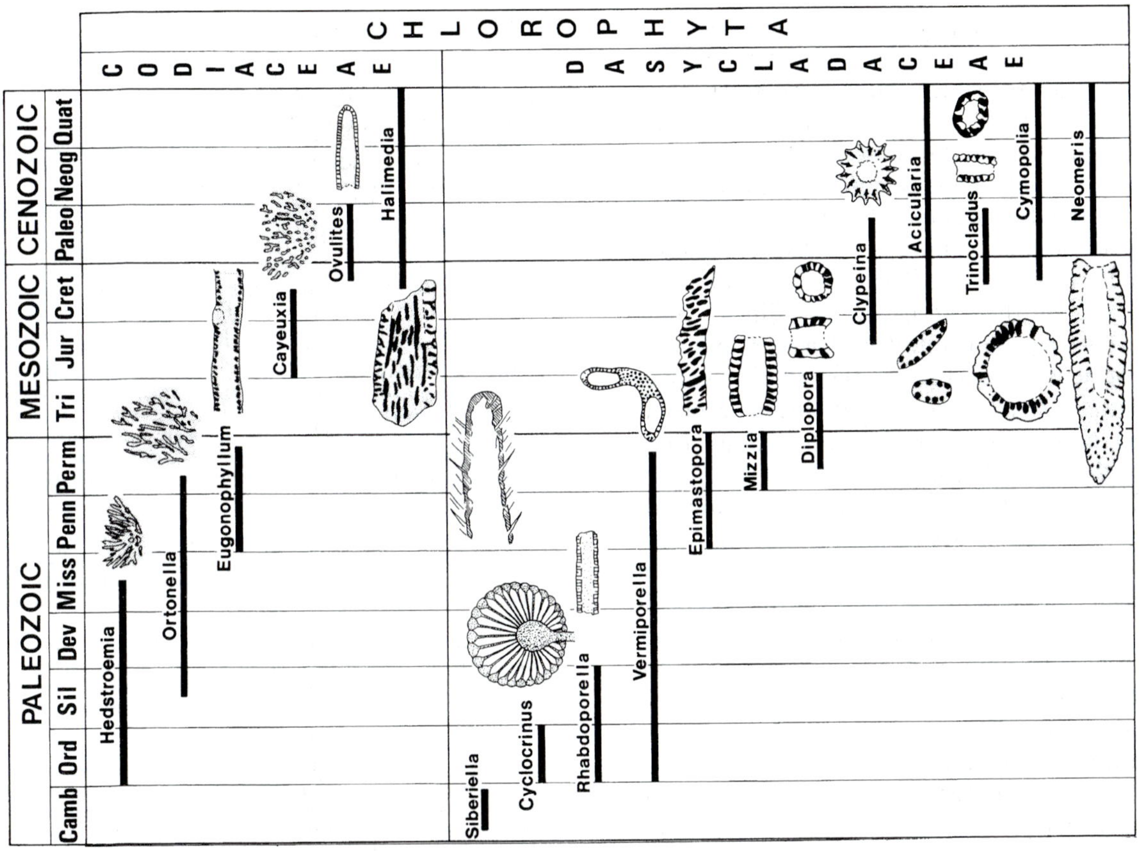

CHLOROPHYTA

CODIACEAE

DASYCLADACEAE

PALEOZOIC | MESOZOIC | CENOZOIC

Camb | Ord | Sil | Dev | Miss | Penn | Perm | Tri | Jur | Cret | Paleo | Neog | Quat

Hedstroemia

Ortonella

Eugonophyllum

Cayeuxia

Ovulites

Halimedia

Siberiella

Cyclocrinus

Rhabdoporella

Vermiporella

Epimastopora

Mizzia

Diplopora

Clypeina

Acicularia

Trinocladus

Cymopolia

Neomeris

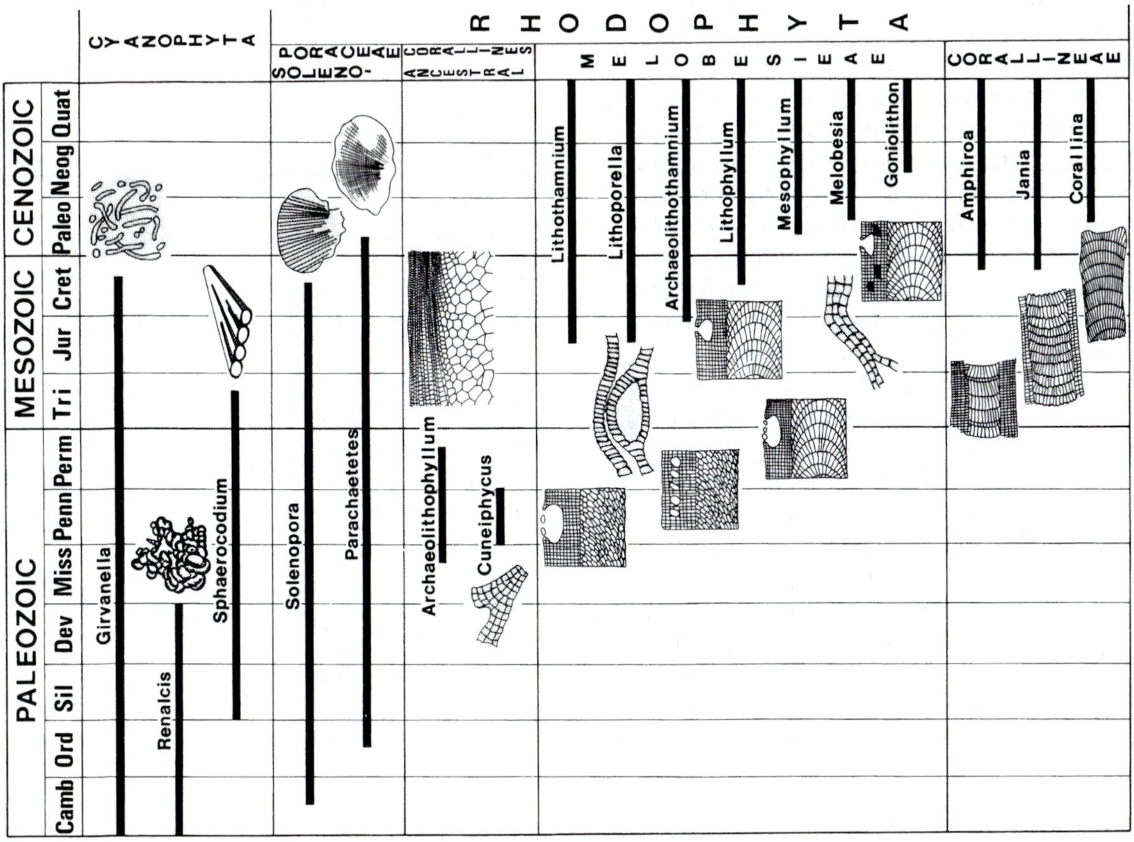

CYANOPHYTA

SOLENOPORACEAE

CORALLINACEAE ANCESTRALS

RHODOPHYTA

MELOBESIEAE

CORALLINEAE

PALEOZOIC | MESOZOIC | CENOZOIC

Camb | Ord | Sil | Dev | Miss | Penn | Perm | Tri | Jur | Cret | Paleo | Neog | Quat

Girvanella

Renalcis

Sphaerocodium

Solenopora

Parachaetetes

Archaeolithophyllum

Cuneiphycus

Lithothamnium

Lithoporella

Archaeolithothamnium

Lithophyllum

Mesophyllum

Melobesia

Goniolithon

Amphiroa

Jania

Corallina

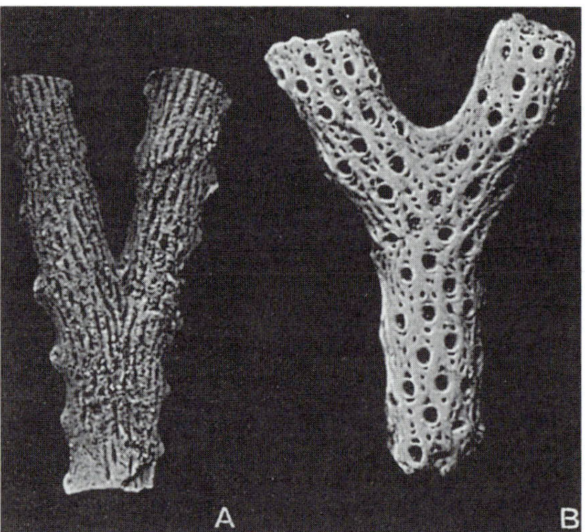

Fig. 24. Stem fragments from frontal (A) and dorsal (B) side of *Hornera striata* Edwards. The genus *Hornera* is typical for the Tertiary deposits and can be recognized by its heavily calcified stems with longitudinal branching ribs which enclose numerous small pores. The apertures of the autozooecia open on one side of the stem only. × 15. Miocene, Italy.

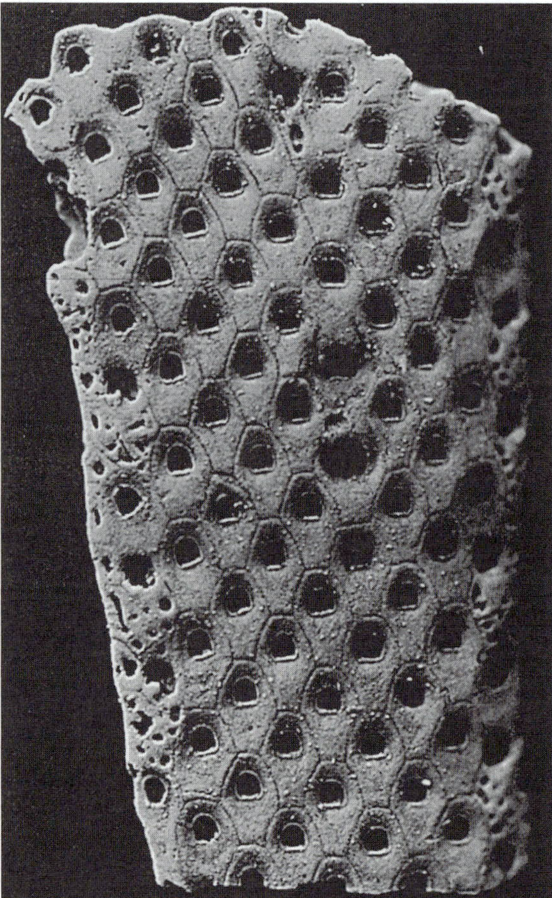

Fig. 25. *Cosinopleura angusta* Berthelsen. The genus *Coscinopleura* is characterized by the porous avicularia located at the lateral sides of the stem. × 20. Danian (Lower Paleocene) of Denmark.

Fig. 26. *Lunulites saltholmiensis* Berthelsen. The genus *Lunulites* is characterized by having a cupuliform, disc-like colony and zooecia arranged in radial rows with interspaced avicularia. × 10. Danian (Lower Paleocene), Denmark.

Fig. 27. *Mucronella hians* Hennig. The genus *Mucronella* is characterized by its clover-leaf-shaped aperture, its almost inperforate zooecium and its small spines around the peristome. × 20. Danian (Lower Paleocene), Denmark.

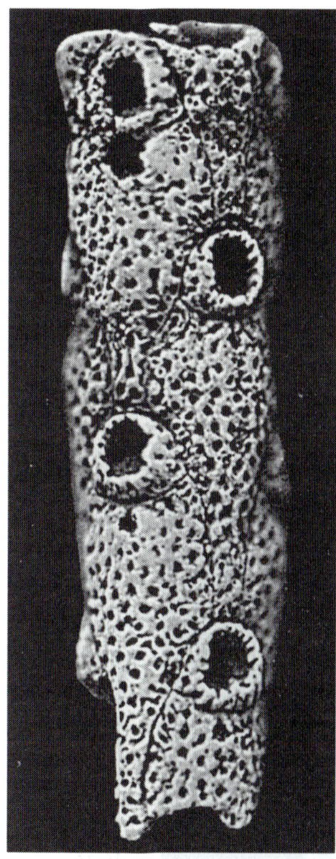

Fig. 28. A *Tubucellaria* species. The genus *Tubucellaria* is recognized by its articulated, cylindrical stems with tubular zooecia and porous zooecial front pierced by an ascopore. × 24. Pliocene of Italy.

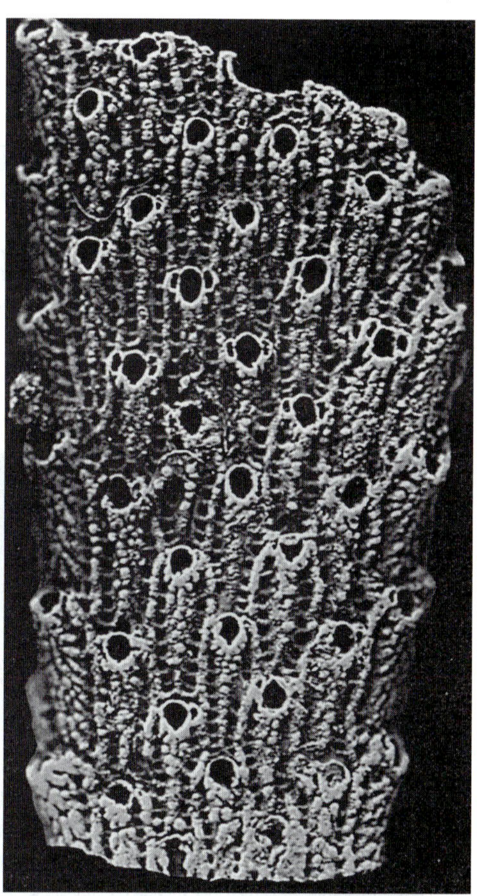

Fig. 29. *Metrarabdotus helveticum* Roger and Buge. The genus *Metrarabdotus* is recognized by its bifoliate colonies with elongated autozooecia with a row of pores at their margin and an avicularium at each side of the aperture. × 20. Miocene, Italy.

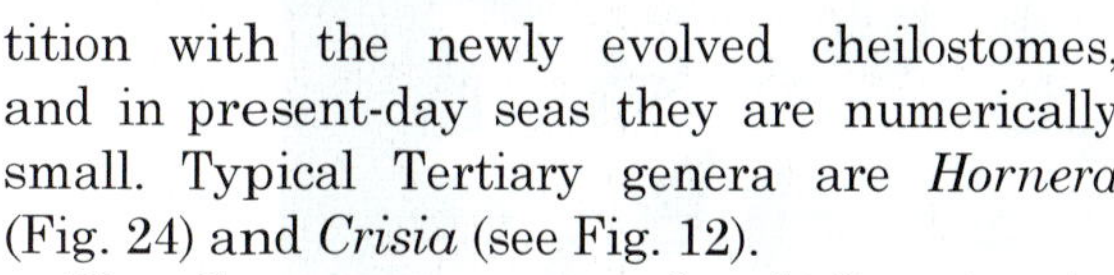

tition with the newly evolved cheilostomes, and in present-day seas they are numerically small. Typical Tertiary genera are *Hornera* (Fig. 24) and *Crisia* (see Fig. 12).

The Ctenostomata are found first in the Upper Ordovician. As this group bears no calcified parts, their fossil record is based on shallow impressions and excavations in shell-bearing organisms.

The cheilostomes evolved from the ctenostomes during the late Jurassic. It was a successful group which developed rapidly. Genera which appear in the Cretaceous are *Onychocella*, *Coscinopleura* (Fig. 25) and *Lunulites* (Fig. 26). The Cheilostomata have dominated the bryozoan fauna since the beginning of the Tertiary. Tertiary genera are *Cellaria*, *Smittipora*, *Membranipora*, *Mucronella* (Fig. 27), *Tubucellaria* (Fig. 28), *Schizoporella* and *Metrarabdotus* (Fig. 29).

SUGGESTIONS FOR FURTHER READING

Bassler, R.S., 1953. In: R.C. Moore (Editor), *Treatise on Invertebrate Paleontology. Part G, Bryozoa.* Geol. Soc. Am., New York, N.Y., and Kansas Univ. Press, Lawrence, Kansas, pp. 1–253. [Contains short descriptions of all the genera known till that time and techniques of sample preparation for Bryozoa.]

Brien, P., 1960. In: P. Grasse (Editor), *Traité de Zoologie*, 5(2): pp. 1054–1335. [General descriptions of the phylum in French.]

Hyman, L., 1959. *The Invertebrates. 5, Smaller Coelomate Groups. Ectoprocta.* McGraw-Hill, New York, N.Y., 619 pp. [General descriptions of the phylum.]

Ryland, J.S., 1970. *Bryozoans.* Hutchinson, London, 175 pp. [An excellent general introduction to the phylum.]

Silén, L., 1966. On the fertilization problems in the gymnolaematous Bryozoa. *Ophelia*, 3: 113–140. [Description of the fertilization process in some Bryozoa.]

Tavener-Smith, R. and Williams, A., 1972. The secretion and structure of the skeleton of living and fossil Bryozoa. *Philos. Trans. R. Soc. Lond.*, 264: 97–159. [Descriptions of the mineralized skeleton and its secretion in Bryozoa.]

CITED REFERENCES

Buge, E., 1957. Les Bryozoaires du Néogene de l'Ouest de la France et leur signification stratigraphique et paléobiologique. *Mem. Mus. Natl. Hist. Nat., Sér. C*, VI: 1–436.

Brood, K., 1972. Cyclostomatous Bryozoa from the Upper Cretaceous and Danian in Scandinavia. *Stockh. Contrib. Geol.*, XXVI: 1–464.

Cheetham, A., 1971. Functional morphology and biofacies distribution of cheilostome Bryozoa in the Danian Stage (Paleocene) of southern Scandinavia. *Smithson. Contrib. Paleobiol.*, 6: 1–85.

Harmelin, J., 1974. *Les Bryozoaires Cyclostomes de Mediterranée. Ecologie et Systematique.* Thesis, University of Aix-Marseille, U.E.R. des Sciences de la Mer et de l'Environment, vol. II.

Wass, R.E., Conolly, J.R. and MacIntyre, R.J., 1970. Bryozoan carbonate sand continuous along southern Australia. *Mar. Geol.*, 9: 63–73.

Basis for subdivision

Families are distinguished in the spumellarians on the basis of overall test shape and wall structure, and in the nassellarians on the basis of homologies in basic structures (Fig. 9). Further subdivision depends generally on structures peculiar to individual family-level taxa with little consistency between them. For example, characters such as pore size, shape and arrangement are commonly significant at the species level, but may characterize even family-level groups.

Note: In the descriptions the following abbreviations are used: S = shape; WS = wall structure; DF = distinctive features; R = remarks; GR = geologic range.

Spumellarians

The spumellarians are distinguished by radial symmetry, and several groups can be readily separated by departure from a strict spherical shape, as well as by wall structure. Thus, the ellipsoidal and discoidal groups involve the lengthening or shortening of one axis, and these groups are further subdivided on the basis of structural peculiarities. Distinctive internal structures in the initial growth stages of several spumellarian groups support close relationships among various discoidal and ellipsoidal groups as suggested by Hollande and Enjumet on the basis of cytological similarities. This leaves a host of spherical forms with widely variable structure, and the systematics of these spumellarians remains one of the most perplexing problems in radiolarian taxonomy.

Entactiniids (Fig. 10)
S: Spherical to ellipsoidal.
WS: Latticed.
DF: Simple, eccentric, internal spicule connected to outer shell by radial bars.
R: Superficially similar internal structures occur in some living actinommids.
GR: Ordovician to Carboniferous.

Orosphaerids (Fig. 11)
S: Spherical to cup-shaped.
WS: Coarse, polygonal, latticed.
DF: Large size (1–2 mm), irregular pore size and shape.
R: Often the only forms preserved in deep sea brown clays (see Friend and Riedel, 1967). May be rare or broken in normal preparations, requiring special separation from coarser size fraction. Fragments of some forms can be identified to genus level for age estimate.
GR: Eocene to Recent.

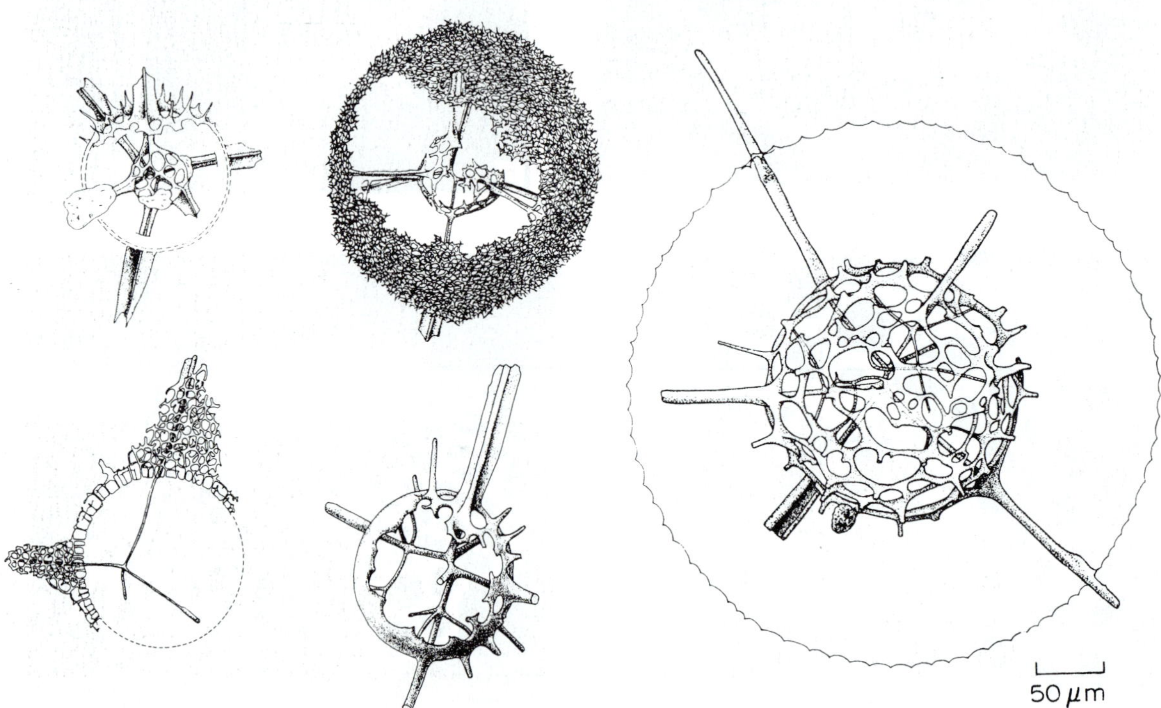

Fig. 10. Entactiniids. Note internal spicule. Drawings reproduced from Foreman (1963) by permission of the author and the American Museum of Natural History, Micropaleontology Press.

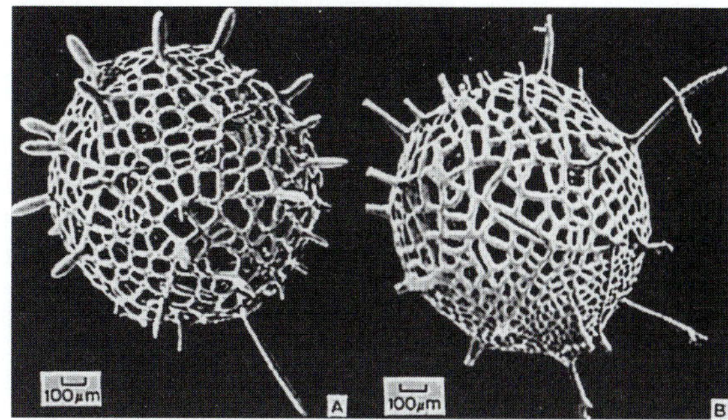

Fig. 11. Orosphaerids. Note coarse, angular lattice.

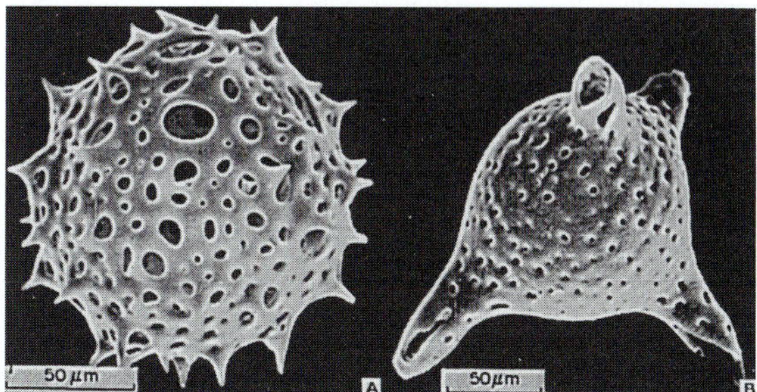

Fig. 12. Collosphaerids.

Collosphaerids (Fig. 12)
S: Spherical to ellipsoidal.
WS: Perforate plate.
DF: Wall structure, thin wall, generally small.
R: Shells often irregular, may bear tubular projections. Colonial forms, colonies reach dimensions of several centimeters, colonies not preserved in sediments.
GR: Lower Miocene to Recent.

Actinommids (Fig. 13)
S: Spherical to ellipsoidal, not discoidal.
WS: Latticed.
DF: Regular latticed wall, little departure from spherical shape, single or multiple shells.
R: A large, polyphyletic family suggested by Riedel for generally spherical forms whose relationships have not yet been determined; thus, subject to future revision and subdivision. Two subfamilies (see below) can be separated at this time.
GR: Paleozoic (?), Triassic to Recent.

Saturnalins (Fig. 14)
S: Spherical.
WS: Latticed.
DF: Outer ring connected to spherical latticed or spongy shell by two or more spines, rarely joins spherical shell directly.
R: Subfamily of actinommids. Forms with spiny outer ring restricted to Mesozoic.
GR: Triassic to Recent.

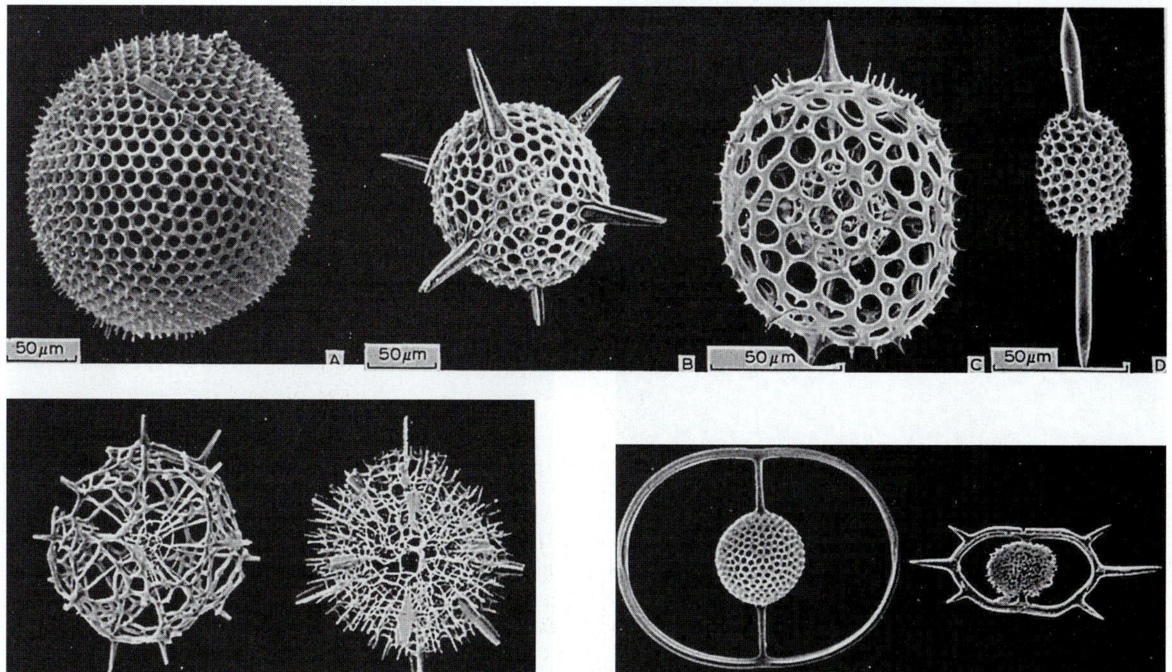

Fig. 13. Actinommids.

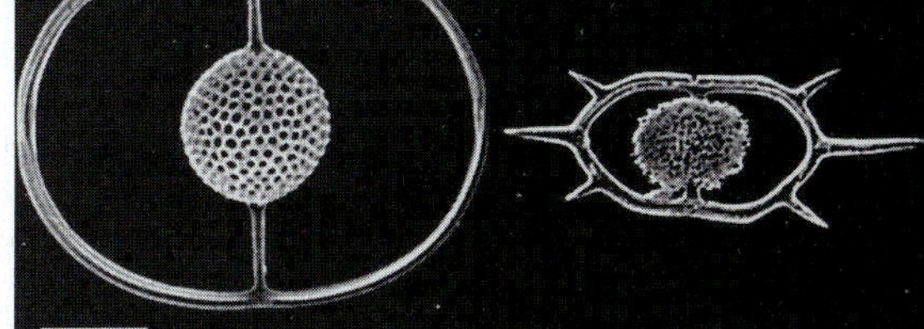

Fig. 14. Saturnalins.

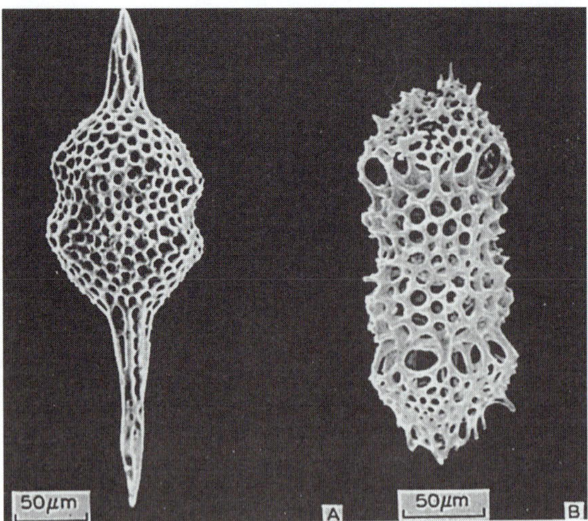

Fig. 15. Artiscins.

Artiscins (Fig. 15)
S: Ellipsoidal.
WS: Latticed, some forms with spongy accessory structures.
DF: Equatorial constriction.
R: Subfamily of actinommids, underwent rapid evolution in Miocene; similar forms, some elliptical with solid spines, others discoidal, are excluded.
GR: Oligocene to Recent.

Phacodiscids (Fig. 16)
S: Discoidal, biconvex to flat, rarely triangular.
WS: Latticed outer shell.
DF: Plain, latticed outer shell separates these from other discoidal families.
GR: Mesozoic (possibly Paleozoic) to Recent.

Coccodiscids (Fig. 17)
S: Discoidal, lenticular.
WS: Latticed.
DF: Latticed shell surrounded by chambered girdles or by chambered or spongy arms.
GR: Mesozoic to Oligocene; doubtful reports from Recent.

Spongodiscids (Fig. 18)
S: Discoidal.
WS: Spongy.
DF: Spongy wall of central disc.
R: A polyphyletic family including some with outer porous plate and some with radiating arms or marginal spines.
GR: Devonian to Recent.

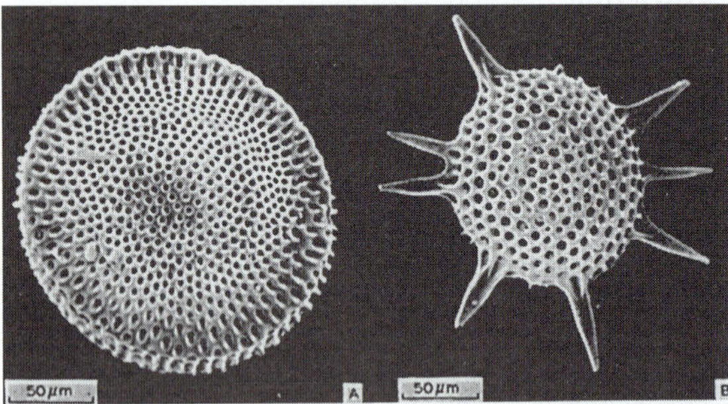

Fig. 16. Phacodiscids.

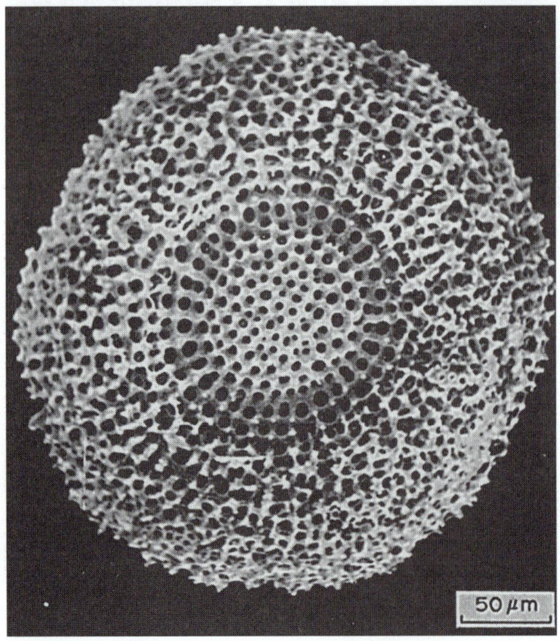

Fig. 17. Coccodiscid.

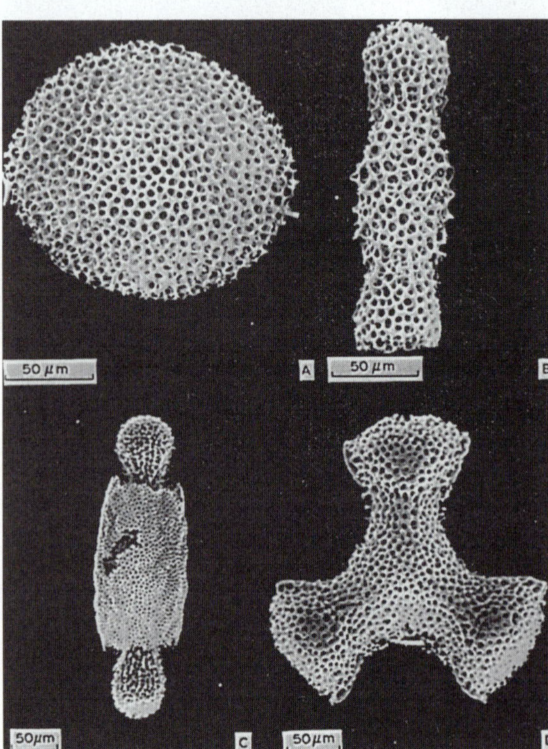

Fig. 18. Spongodiscids.

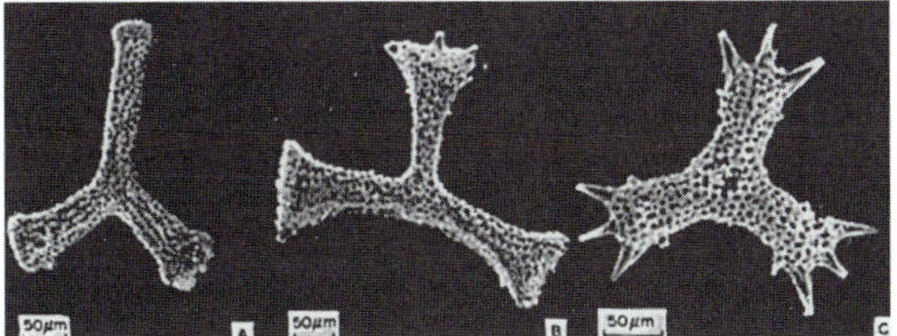

Fig. 19. Hagiastrids. Reproduced from Pessagno (1971) by permission of the author and the Paleontological Research Institution.

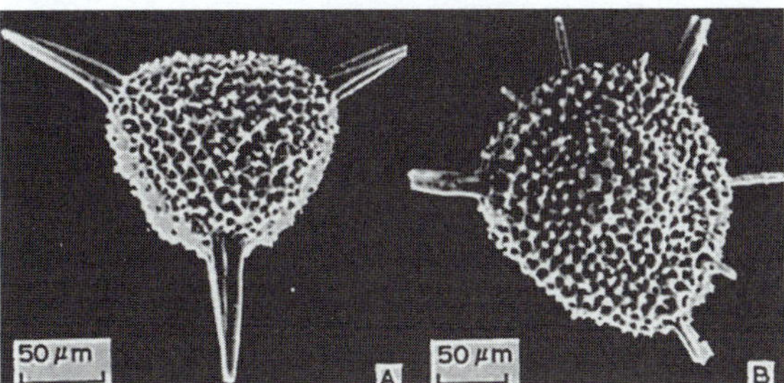

Fig. 20. Pseudoaulophacids. Reproduced from Pessagno (1972) by permission of the author and the Paleontological Institution.

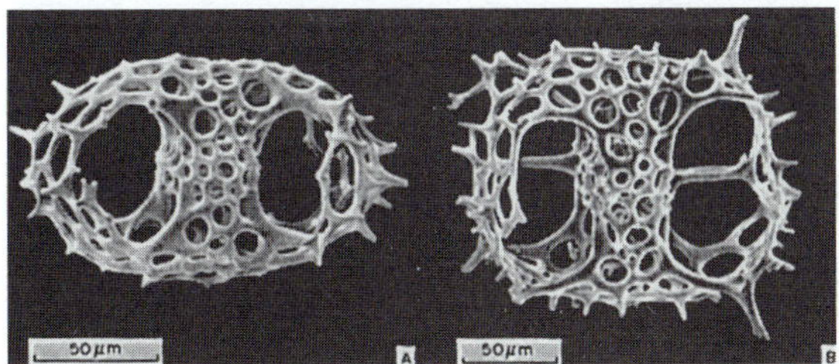

Fig. 21. Pyloniids.

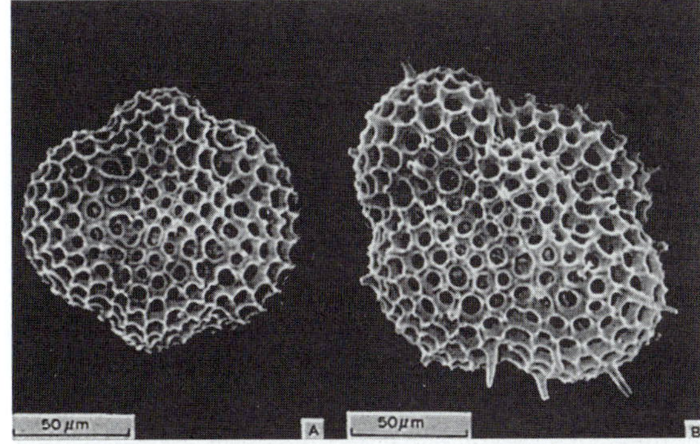

Fig. 22. Tholoniids.

Hagiastrids (Fig. 19)
S: Discoidal, flat.
WS: Spongy, rectangular.
DF: Two, three or four radial arms of regular, rectangular spongy meshwork.
R: Radial arms usually more conspicuous than central area.
GR: Mesozoic.

Pseudoaulophacids (Fig. 20)
S: Discoidal, lenticular to triangular.
WS: Spongy, triangular.
DF: Regular, equilaterial triangular meshwork in concentric layers.
R: Usually bear a few prominent marginal spines.
GR: Mesozoic.

Pyloniids (Fig. 21)
S: Ellipsoidal.
WS: Latticed.
DF: Successively larger, latticed elliptical girdles in three perpendicular planes.
GR: Eocene to Recent; common only from Miocene to Recent.

Tholoniids (Fig. 22)
S: Ellipsoidal.
WS: Latticed.
DF: Cortical shell divided into dome-shaped segments separated by annular constrictions or furrows.
R: Initial chamber of basic pylonid structure; rare forms, few species.
GR: Pliocene to Recent.

Litheliids (Fig. 23)
S: Ellipsoidal (rarely spherical) to lenticular.
WS: Latticed.
DF: Internal structure coiled.
R: Initial chamber of basic pyloniid structure.
GR: Carboniferous to Recent.

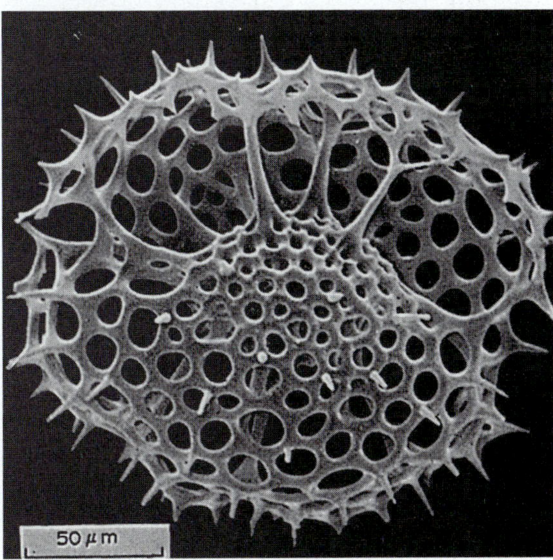

Fig. 23. Litheliid. Note coiling.

Nassellarians

Nassellarians are subdivided on the basis of homologies in a basic skeletal element (see Fig. 9). A prominent apical spine is often termed apical horn, while prominent dorsal and lateral spines may be termed feet. The median bar, apical, dorsal and primary lateral spines, or their homologues are almost invariably recognizable by their relative sizes and angular relationships. This simple spicule itself is the basis for one family, while another is based on a D-shaped ring (Fig. 9D), either isolated or as a prominent sagittal ring (Fig. 9C), resulting from an arched connection between the apical and vertical spines. In other forms, the basic spines are joined together in a latticed chamber called the cephalis, whose size, shape and structure characterize families. In multi-chambered forms, the first two post-cephalic segments are termed **thorax** and **abdomen**.

Plagoniids (Fig. 24)
S: Simple nassellarian spicule or single latticed chamber (cephalis).
DF: Basic spicule without post-cephalic chambers.
R: Wide variety of forms developed from accessory spines and branches, including latticed chamber surrounding spicule; probably a polyphyletic group subject to future subdivision.
GR: Cretaceous to Recent.

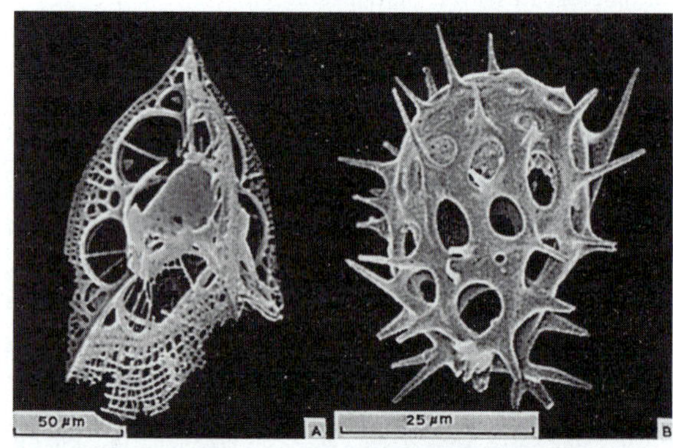

Fig. 24. Plagoniids.

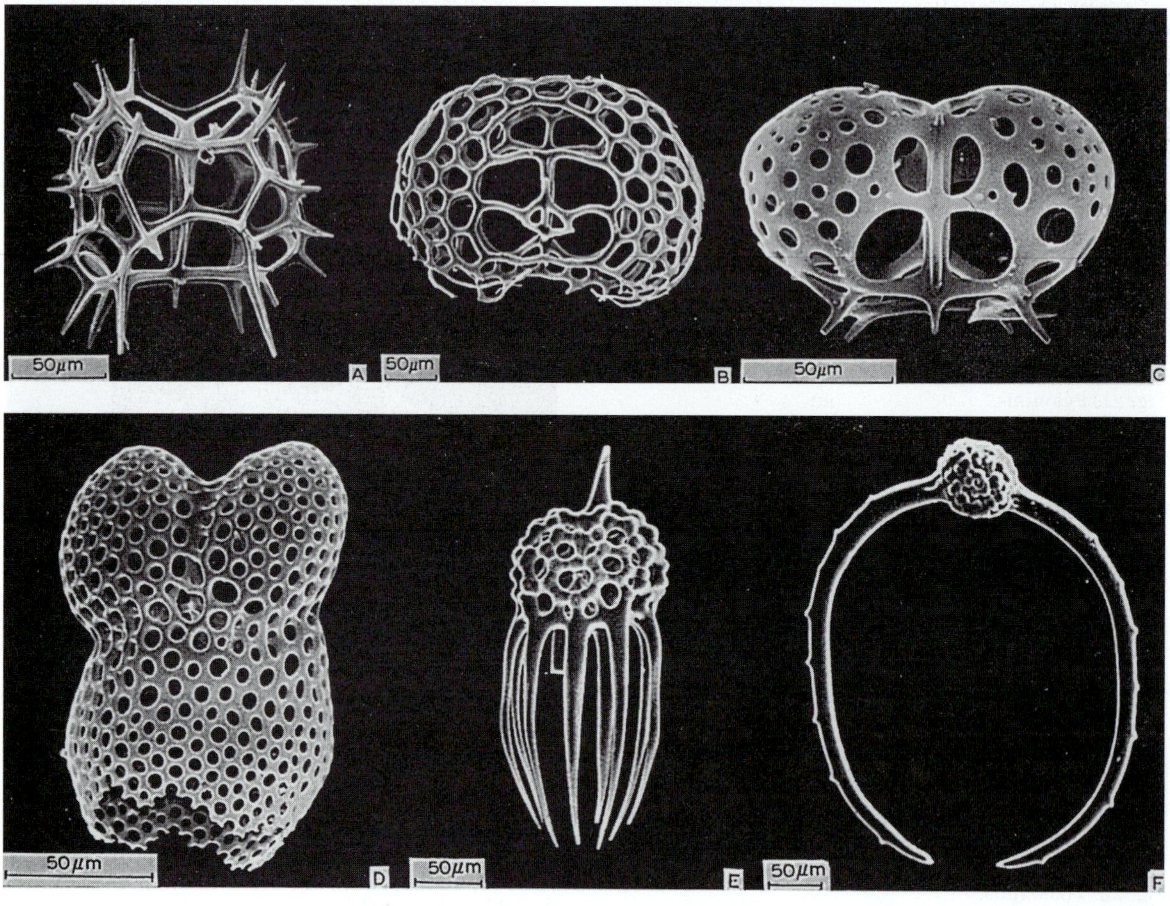

Fig. 25. Acanthodesmiids.

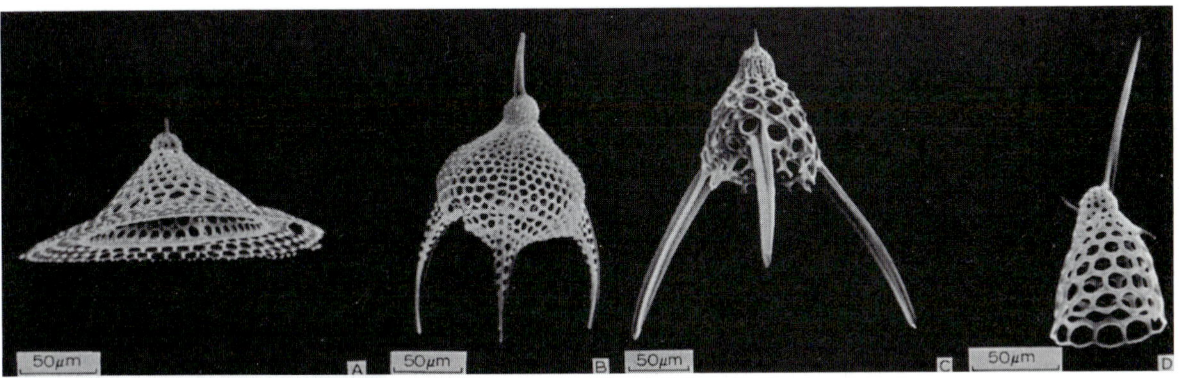

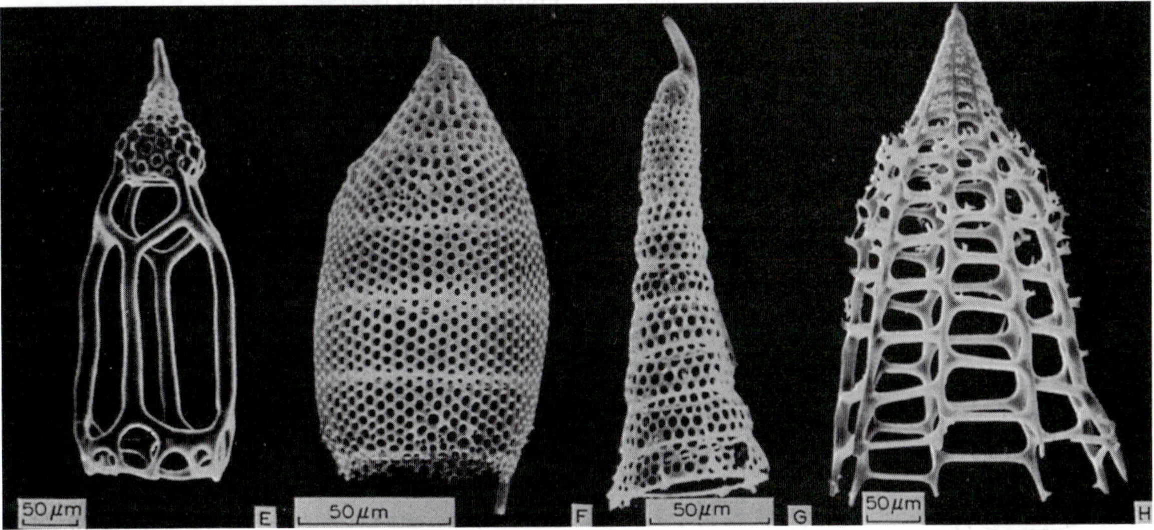

Fig. 26. Theoperids.

Acanthodesmiids (Fig. 25)
S: D-shaped ring or latticed, bilobed chamber with D-shaped sagittal ring.
DF: D-shaped ring always conspicuous externally.
R: Forms range from simple rings to latticed chambers consisting of lobes developed on either side of the D-ring; family has been revised by Goll (1968, 1969) under name Trissocyclidae.
GR: Cenozoic.

Theoperids (Fig. 26)
S: Small spherical cephalis and one or more post-cephalic chambers.
DF: Cephalis usually poreless or sparsely perforate.
R: Cephalis contains reduced internal spicule homologous with that of plagoniids; a large, probably polyphyletic group containing majority of ordinary cap- or helmet-shaped nassellarians.
GR: Triassic to Recent.

Carpocaniids (Fig. 27)
S: Small cephalis merging with thorax.
DF: Cephalis nearly indistinguishable from thorax, often reduced to a few bars that are homologous with spicule in other groups.
GR: Eocene to Recent.

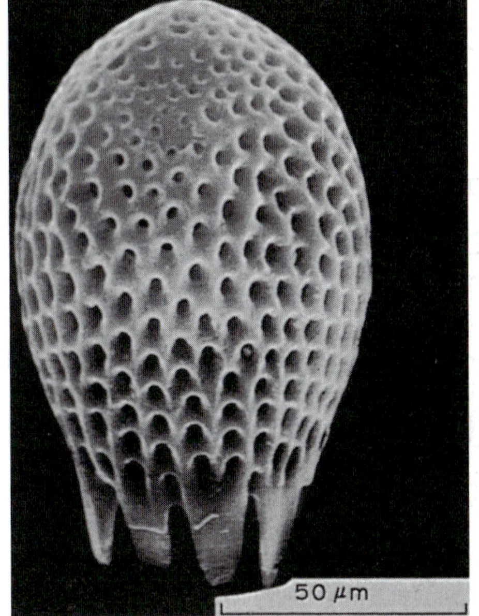

Fig. 27. Carpocaniid.

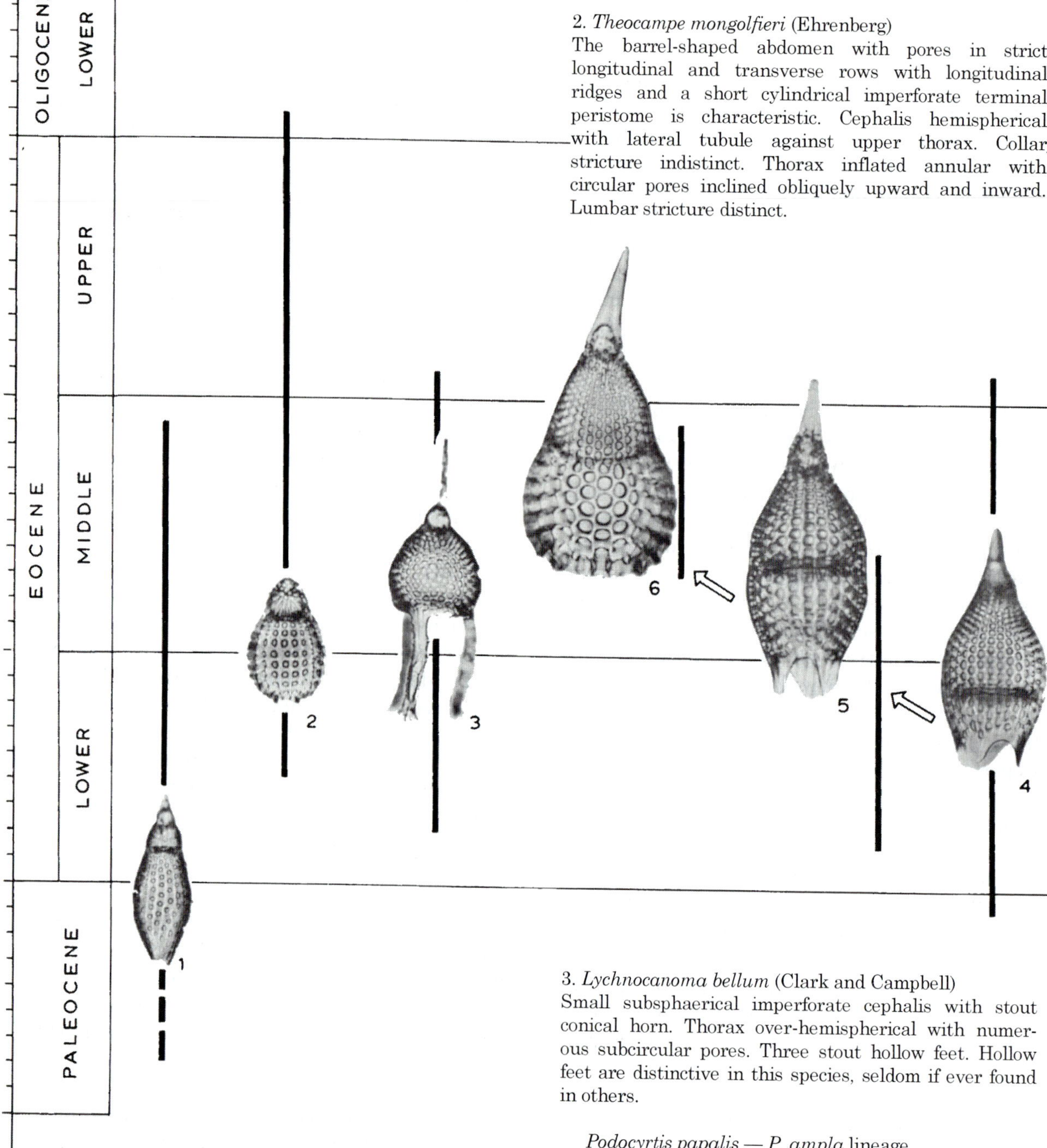

1. *Phormocyrtis striata* Brandt
Characterized by overall fusiform shape. Cephalis hemispherical, with small pores and bladed horn. Collar stricture indistinct. Thorax hemispherical with irregularly arranged circular pores. Lumbar stricture indistinct. Abdomen fusiform with greatest width near middle, with circular pores in longitudinal rows usually separated by ridges. Termination ragged, toothed, or rarely closed. Paleocene and early Eocene specimens with abdomen usually triangular in cross-section.

2. *Theocampe mongolfieri* (Ehrenberg)
The barrel-shaped abdomen with pores in strict longitudinal and transverse rows with longitudinal ridges and a short cylindrical imperforate terminal peristome is characteristic. Cephalis hemispherical with lateral tubule against upper thorax. Collar stricture indistinct. Thorax inflated annular with circular pores inclined obliquely upward and inward. Lumbar stricture distinct.

3. *Lychnocanoma bellum* (Clark and Campbell)
Small subsphaerical imperforate cephalis with stout conical horn. Thorax over-hemispherical with numerous subcircular pores. Three stout hollow feet. Hollow feet are distinctive in this species, seldom if ever found in others.

Podocyrtis papalis — P. ampla lineage

This lineage consists of three broadly defined species.

4. *Podocyrtis papalis* Ehrenberg
Diagnostic features are the inflated conical thorax passing into inverted truncate-conical abdomen without external expression of the lumbar stricture, the longitudinal rows of pores separated by ribs, and the pored part of the abdomen shorter than the thorax and below that a poreless part with three large, shovel-shaped feet.

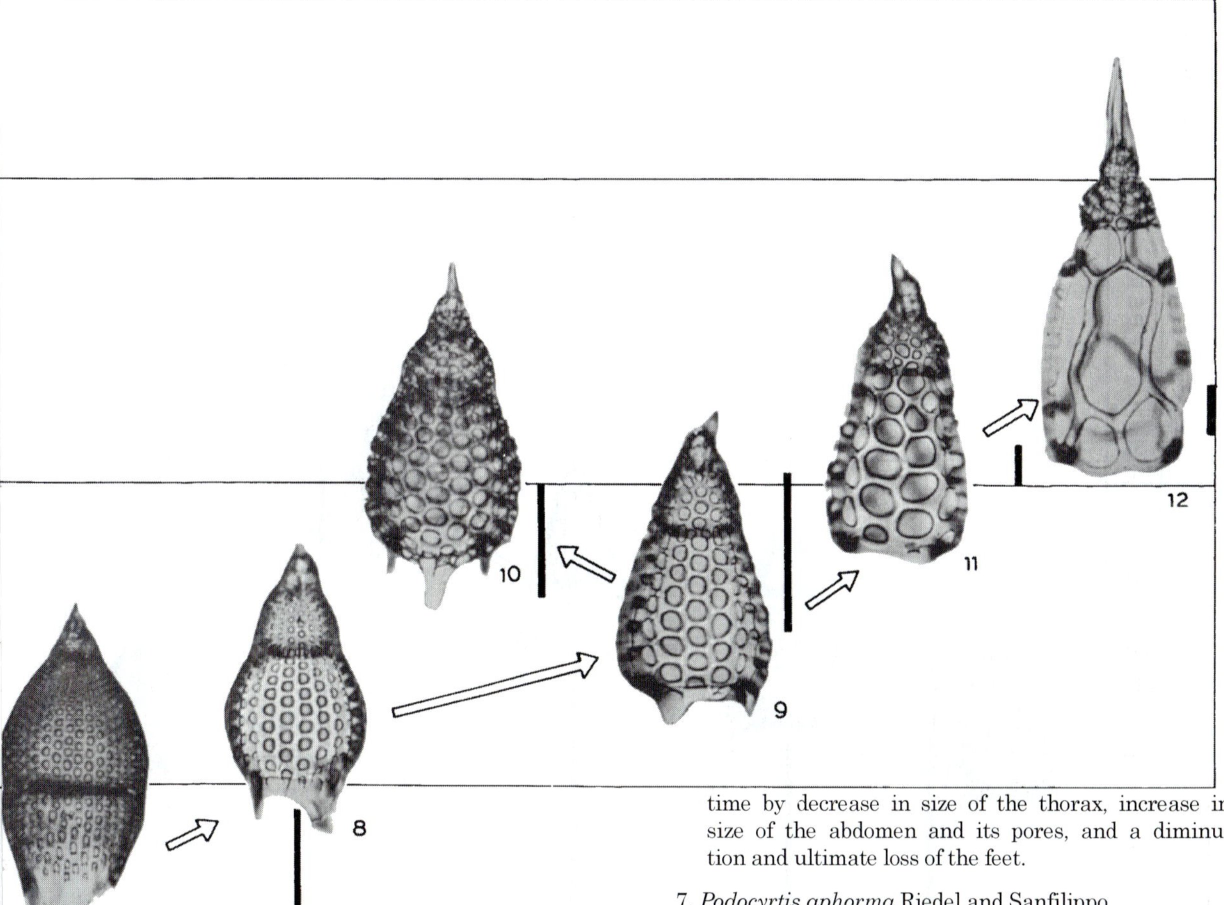

5. *Podocyrtis diamesa* Riedel and Sanfilippo
A form intermediate between *P. papalis* and *P. ampla*, differing from *P. papalis* by its larger size and the presence of a distinct lumbar stricture, and from *P. ampla* in its general spindle-shaped rather than conical form. Thorax and abdomen usually of approximately same length. Pores separated by ridges in early specimens, larger and lacking intervening ridges in later specimens. Three feet shovel-shaped, irregular in some specimens with very restricted apertures.

6. *Podocyrtis ampla* Ehrenberg
Characterized by conical overall shape and abdomen terminating in narrow thickened rim bearing three small shovel-shaped or spathulate feet. Collar and lumbar strictures not pronounced. Abdominal pores larger than thoracic, usually in longitudinal rows without intervening ridges.

Podocyrtis aphorma — P. goetheana lineage
P. aphorma evolved from *P. papalis* and then developed through three intermediate forms to *P. goetheana*. *P. trachodes* is apparently a related side branch. The succession is characterized with time by decrease in size of the thorax, increase in size of the abdomen and its pores, and a diminution and ultimate loss of the feet.

7. *Podocyrtis aphorma* Riedel and Sanfilippo
Distinguished from *P. papalis* only by less regular abdominal pores and the presence of a slight lumbar stricture.

8. *Podocyrtis sinuosa* Ehrenberg (?)
Distinguished from *P. aphorma* by the larger abdomen and from *P. trachodes* by the smoother surface. With time, the abdomen increases in size while the thorax decreases.

9. *Podocyrtis mitra* Ehrenberg
Distinguished from *P. sinuosa* and *P. trachodes* in having its abdomen widest near the distal end, rather than medially.

10. *Podocyrtis trachodes* Riedel and Sanfilippo
Distinguished by the rough surface of its thorax and (especially) abdomen.

11. *Podocyrtis chalara* Riedel and Sanfilippo
Distinguished from *P. mitra* by larger abdominal pores and in generally lacking feet.

12. *Podocyrtis goetheana* (Haeckel)
Distinguished from *P. chalara* by even larger abdominal pores, a medial transverse row of which are distinctly elongate.

such hyaline regions run down the center of the valve parallel to the apical axis and are called the **pseudoraphe**. Some pennate genera have a **V**-shaped slit, **raphe**, in place of the pseudoraphe.

Additional structures, although seemingly insignificant, are important even at the generic level. A small hyaline area, **pseudonodule**, near the periphery of centric diatoms, for example, identifies the genus *Actinocyclus* while two or more pores near the central area characterize the genus *Thalassiosira*, and a circlet of spines identifies the genus *Stephanopyxis*. It is essential, therefore, to emphasize that a careful scrutiny of the diatom valve is necessary for accurate specific identification.

Reproduction

Diatoms reproduce by simple cell division. Just prior to division, the epitheca and hypotheca move slightly away from each other and separation of the two valves takes place (see Fig. 1). New valves are formed on the exposed protoplasm from the central area outwards and always within the confines of the parent valve. Thus, the original two valves both become epitheca in the new individuals. One can project, therefore, a steady decrease in valve diameter with each succeeding generation. Not only do the valves become smaller, but they also change in geometry as well as in the character of their surface markings.

In some reproductive phases, the diatom will subdivide to a very small size and to a low level of vitality before dying. In others, cell division will cease at some level and a special cell, called an **auxospore**, will be formed. The purpose of this auxospore, although the mechanics are not well understood in most species, is to return the diatom species to its original size from which the process starts anew.

Just as important to species survival as the reproductive cycle is the ability of many diatoms to survive from one growing season to another. Some species will persist through the winter months, their life processes greatly reduced. Most, however, form a heavy resting spore which falls to the bottom and is revived when conditions favorable for growth return.

Nutrition

The nutrient content of the water is extremely important to diatom growth and reproduction. Three nutrients are considered essential for almost all diatom species. These are phosphorus, nitrate and silica. The Russian worker T.V. Belayeva has reported that the giant marine diatom *Ethmodiscus rex* (Rattray) Hendey lives best in waters of low phosphate concentration, but this preference is considered atypical for diatom species. Experience, both in the laboratory and in the field, has shown that when any of these three basic nutrients is missing, diatom growth and reproduction ceases. Some diatoms are so sensitive that even a slight reduction in the supply of these dissolved nutrients will inhibit their reproduction.

Given these facts it is relatively easy to point out areas of high diatom productivity. Any area of upwelling in the ocean will bring a constant supply of these nutrients to the surface and, thus, cause productivity blooms. Coastal regions which receive high concentrations of nutrients from both run-off from land and rain will support large diatom populations.

In addition to the nutrient supply many species have special requirements. A number, for example, are known to require cobalamin (vitamin B_{12}) as well as thiamin (vitamin B_1). Sulphur, iron and manganese are also considered essential for many diatoms. In addition, various diatom species are known to respond to trace elements in the water. Thus, a number of nutritional variables may be operating to promote or inhibit the presence of specific diatom species.

ECOLOGY

Habitats

Within both fresh-water and marine environments diatoms can be found occupying a great number of niches. On land they are found in soils and occasionally on wetted rocks and plants. In streams, lakes and ponds they are found attached to rocks and plants as well as in bottom muds.

In the marine realm, two broad diatom

habitats are recognized. Benthic diatoms live in littoral environments either attached (**sessile**) or capable of some movement along the bottom (**vagile**). They are of considerable importance in that they serve as the major food for many shallow-water marine feeders. In addition, certain species produce enough cementing mucus to cement sediment grains together and thus inhibit submarine erosion. Other diatom species are known to form a dense mat at the sediment/water interface and thus prevent any continued reworking of the sediment.

Planktonic species are considered as part of the oceanic or neritic plankton. Oceanic (holoplanktonic) forms spend their entire existence in the open ocean and pass through the various phases of their life cycle in that environment. Neritic forms are found in association with coastlines. Since such forms frequently pass through a benthic stage, these coastal regions must possess rather shallow shelves.

Hendey (1964) divides neritic plankton into three categories: holoplanktonic, meroplanktonic and tychopelagic. The **holoplanktonic** existence was considered previously and is mentioned here only because some neritic plankton appear to be holoplanktonic. **Meroplanktonic** species live close to the coastline in the plankton but spend part of their existence in bottom sediments, probably as resting spores. It is sometimes difficult to differentiate holoplanktonic and meroplanktonic species. **Tychopelagic** species probably spend most of their lives on the bottom (Fig. 5).

Frequently a diatom assemblage far from the present coastline may contain representatives of both the neritic and oceanic plankton as well as a number of sessile and vagile forms. This is understandable since bottom currents or storms may dislodge shallow-water forms and transport them many kilometers from their original habitat.

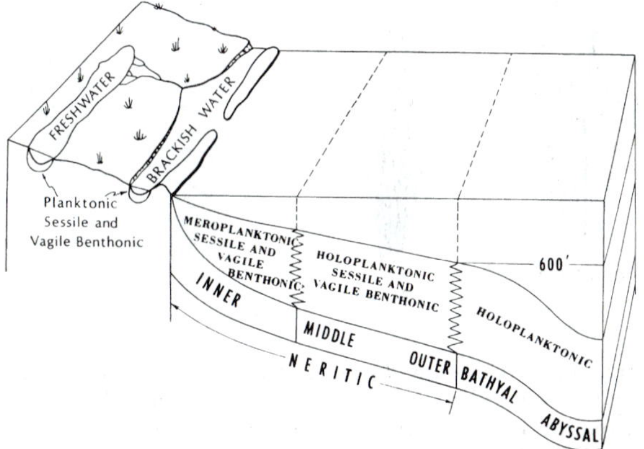

Fig. 5. Marine and fresh-water environments occupied by diatoms. (From Wornardt, 1969).

Factors affecting distribution

Factors which control diatom distribution vary with habitat. Temperature and salinity appear to be the principal factors controlling distribution of marine diatoms. Thus, diatoms are relatively easy to culture, although frequently it is difficult to reconstruct diatom paleoecology.

MAJOR MORPHOLOGICAL GROUPS

Diatoms take on a great many shapes and thus it is difficult to present a simple picture of the major morphological groups. Most diatomists follow the lead of their nineteenth century predecessors in recognizing two major divisions — the Centrales and the Pennales. The Centrales may be circular, oblong, hemicircular, triangular, or quadrangular, but the surface structures are arranged with reference to some central or near-central point. The Pennales, on the other hand, are elongate with major structures at approximately right angles from a median line which runs parallel to the long axis.

The Pennales can further be divided into those forms which possess a true raphe, or cleft, and those that do not. Any further morphological breakdown is largely based on such features as shell geometry and surface structures. Although the many different kinds of shapes, geometries and structures in the diatom present a somewhat confusing picture, N. Ingram Hendey (1964) has succeeded in assigning most diatoms to one of seven "shape groups". These seven minor groups can be lumped together into two major groups: (1) those whose valves usually have a raphe or pseudoraphe; and (2) those whose valves are without a raphe or pseudoraphe. In the first category, Hendey recognized four "shape groups": the linear diatoms, the cuneate diatoms, the cymbiform diatoms and the carinoid diatoms.

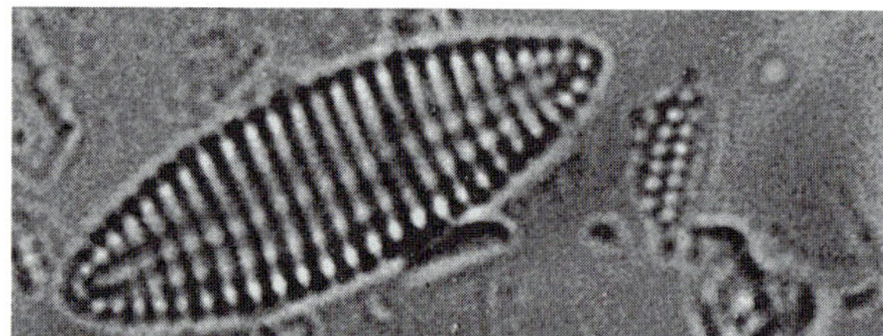

Nitzschia miocenica Burckle. Valve flat, linear-lanceolate with gently rounded apices. Two lines run down center of valve, some 2–3 μm apart. Intercostal membrane with two rows of small punctae. × 1000.

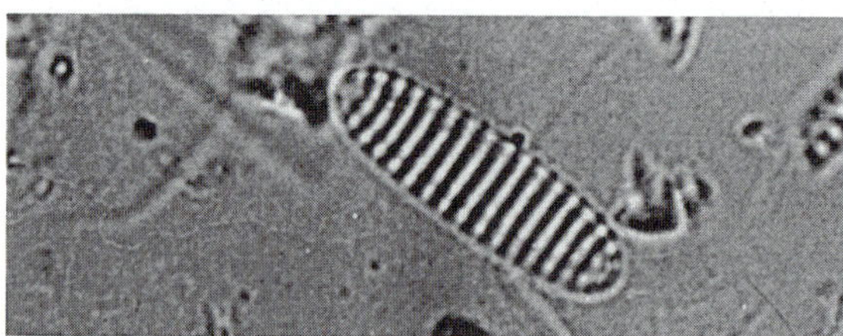

Nitzschia porteri Frenguelli. Valve small linear-elliptical with broadly rounded ends. Costae close together and intercostal membrane with rows of small punctae. × 1000.

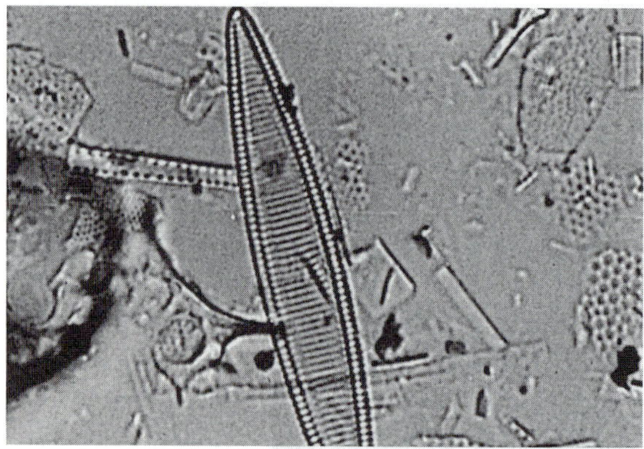

Nitzschia reinholdii Kanaya and Koizumi. Valve elliptical with slightly convex margins. Costae close together. Intercostal membrane with rows of small decussate punctae. × 1000.

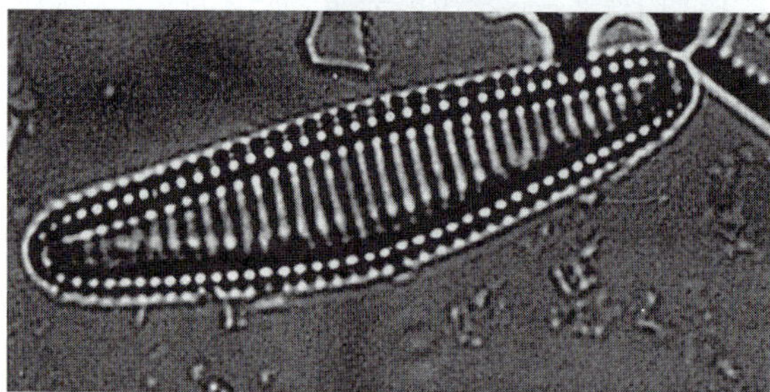

Pseudoeunotia doliolus (Wallich) Grunow. Valve heteropolar with dorsal margin gently convex and ventral margin gently concave. Costae close together. Intercostal membrane with rows of small decussate punctae. × 1000.

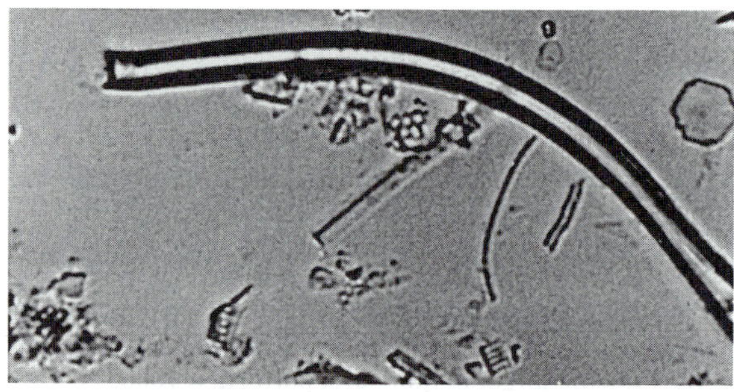

Rhizosolenia barboi Brun. Curving cylindrical tube terminated at the apical end by two small spines. Spine is present at point of maximum curvature. × 1000.

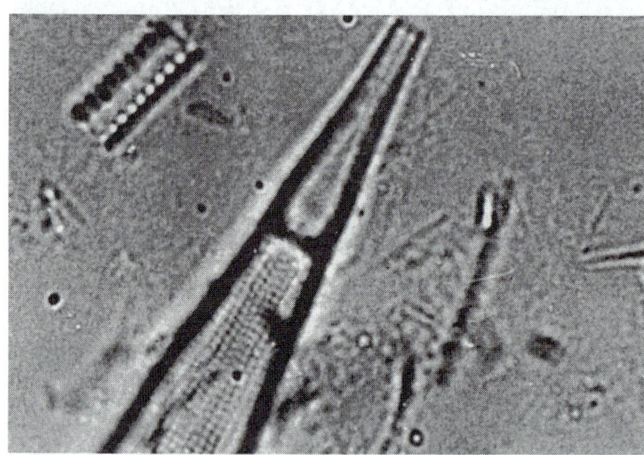

Rhizosolenia bergonii Peragallo. Gently tapering tube terminating at the apical process. Calyptra is finely punctate. × 1000.

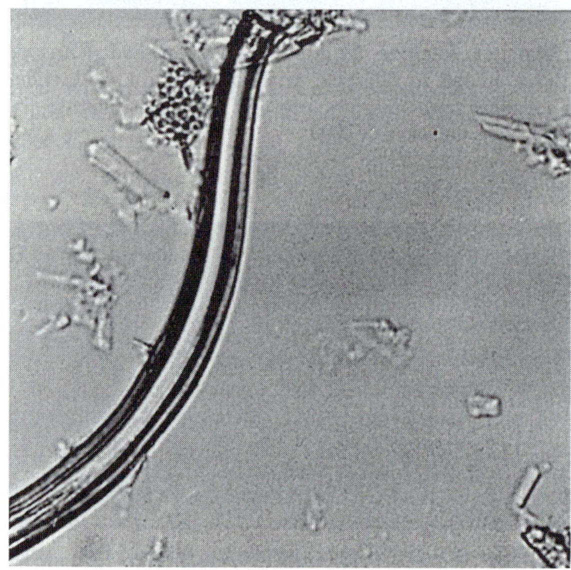

Rhizosolenia curvirostris Jousé. Similar to *R. barboi* except that it is larger and lacks the spine at the point of maximum curvature. × 1000.

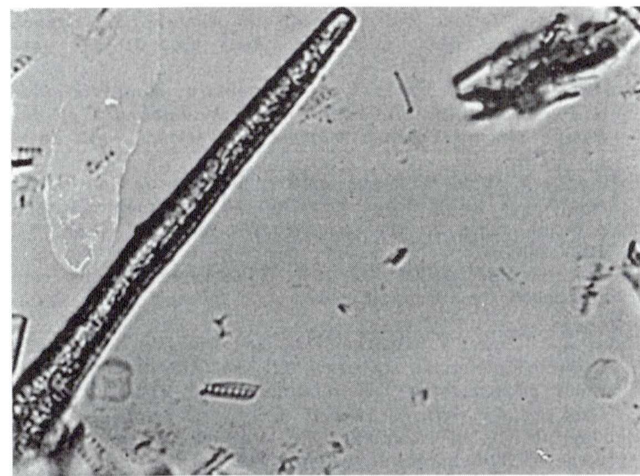

Rhizosolenia praealata Schrader. Valve cylindrical, with rows of radial punctae extending up to the middle of the apical process. Apical process is without spines. × 1000.

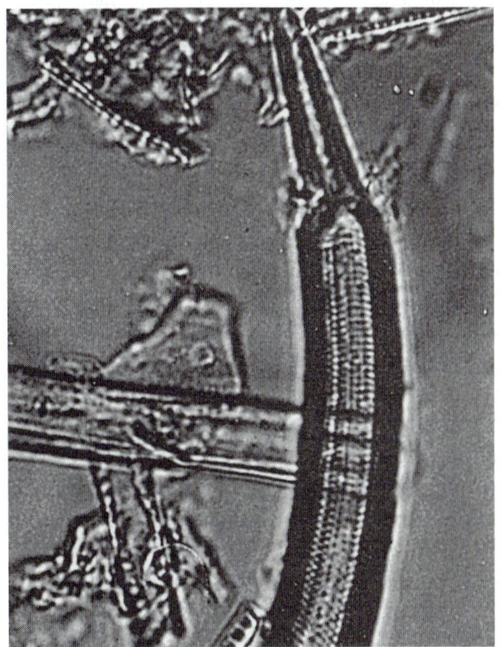

Rhizosolenia miocenica Schrader. Valve cylindrical, tapering toward the apex. Apical process with lines of punctae. Apical spine present. × 750.

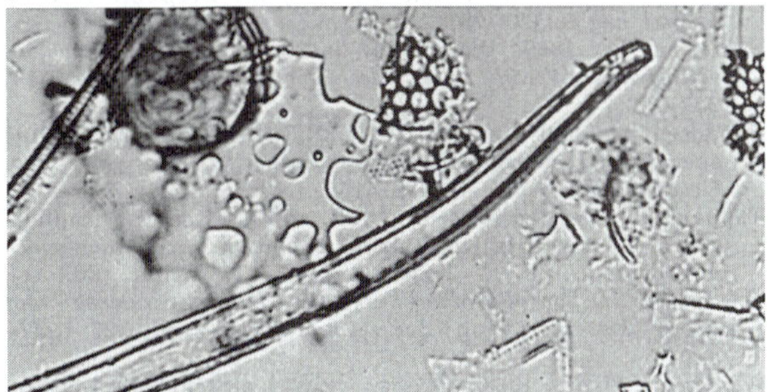

Rhizosolenia praebarboi Schrader. Valve cylindrical, straight or gently curved. Apical process hyaline. No apical spine. × 750.

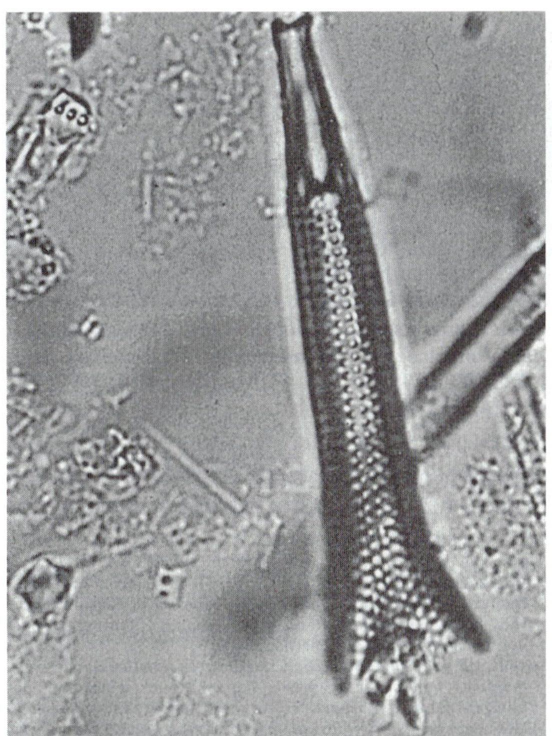

Rhizosolenia praebergonii Muchina. Valve cylindrical, robust, strongly areolate. Apical process thick, flaring out at the apex. × 1000.

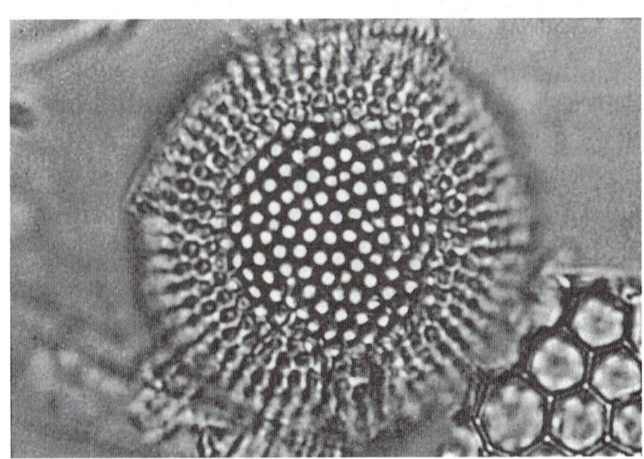

Thalassiosira convexa Muchina. Valve circular, strongly convex. Areolae coarse, frequently fasciculate. Well-developed margin. × 1000.

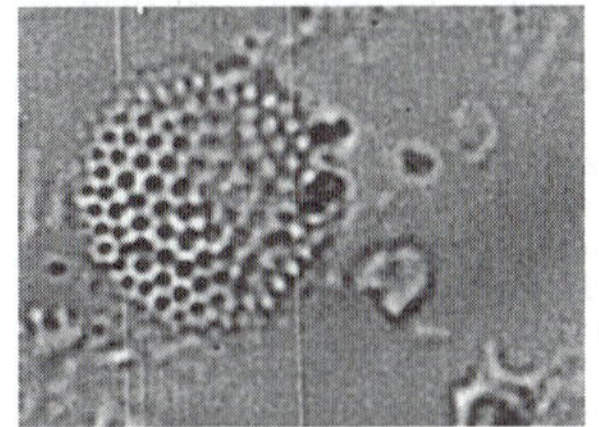

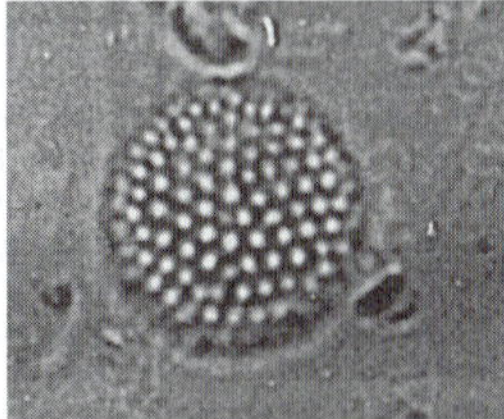

Thalassiosira praeconvexa Burckle. Valve small, convex. Fine areolae are frequently in fascicules. × 1000.

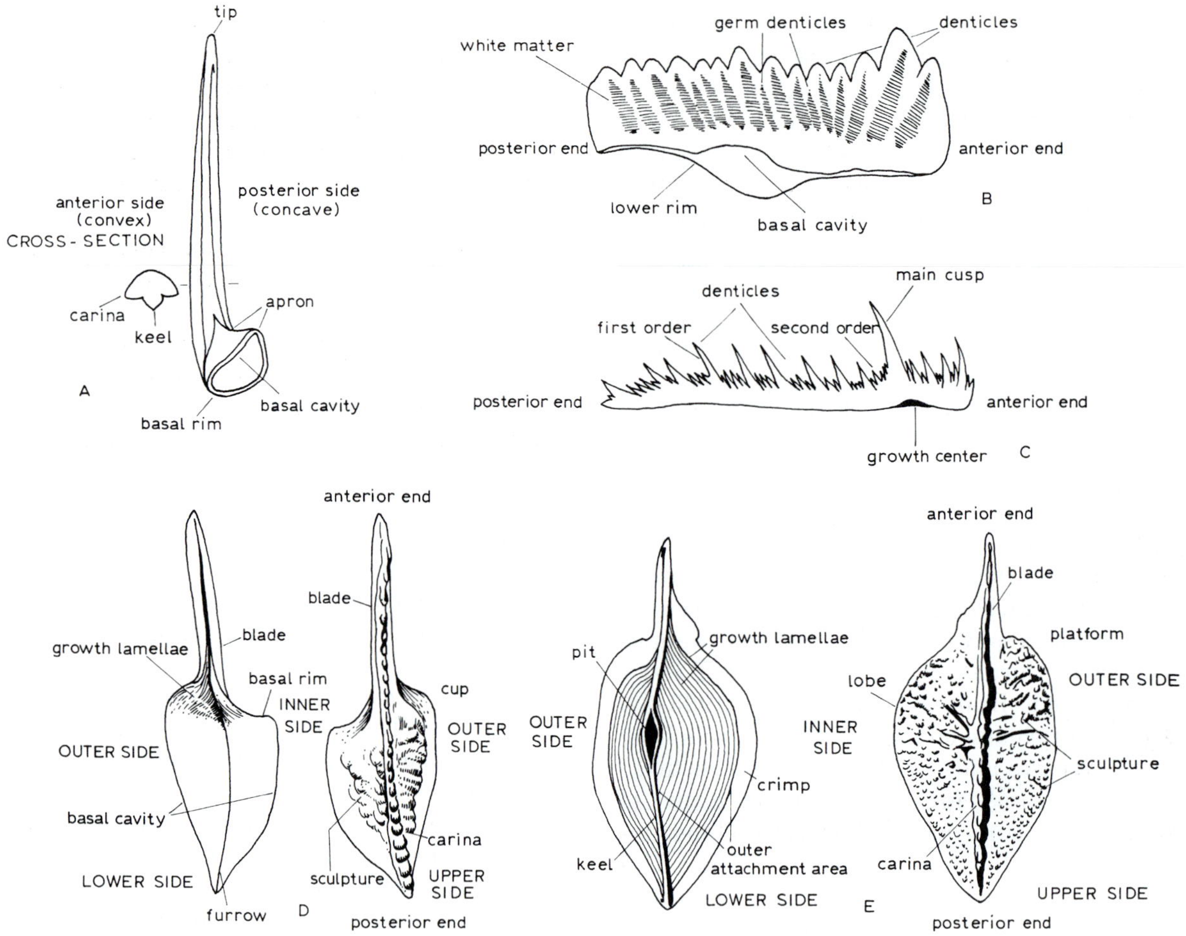

Fig. 5. Morphological terminology of conodonts. A. Single-cone type. B. Blade type. C. Bar type. D. Platform type (*Gnathodus*). E. Platform type (*Polygnathus*).

and development of **keels** and **carinae** on the sides are of systematic value. They are observed in cross-sections of denticles above the **apron**. The flaring portion above the **basal rim**, and its development and shape is also of taxonomic importance.

Germ denticles is a term for structures which have been suppressed during ontogeny due to overgrowth by adjacent structures. They can be observed in translucent fossils in transmitted light.

In platform-type conodonts a **blade** is the edge-like free portion at the anterior end. In many cases it integrates into a carina on the upper side of the platform. Protrusions on one or both sides of the platform are named **lobes**.

On the lower side of platforms, such as in *Polygnathus* (see Fig. 5) the main elements are: (1) the **pit** around the growth center; (2) the attachment area where the edges of **growth lamellae** are visible and to which the basal organ is attached; (3) the **crimp**, or the portion outside the basal organ which does not expose the edges of growth lamellae; and (4) the keel in the center, which corresponds to the carina at the upper surface.

The lower side of some platforms, such as *Gnathodus* (see Fig. 5) is entirely excavated, and exposes the edges of growth lamellae on the entire surface. It is termed a **cup**. In the center a furrow, which is homologous to the keel, is developed.

Internal structures

The study of internal structures is important for suprageneric systematics as well as for investigations of the nature and function of conodonts. Initially, a nucleus was formed by a lamella. In subsequent stages this center is covered from outside by growth lamellae. In the earliest conodonts (suborder Paraconodontida), the growth lamellae are always open at the "upper side" (Fig. 6). In the suborder Conodontiformes, the upper lamellae are uninterrupted and at the lower side of the conodont a basal plate which is composed of much finer crystallites is developed (Fig. 7). It is not as sturdily built as the rest of the conodont and in many instances is not preserved.

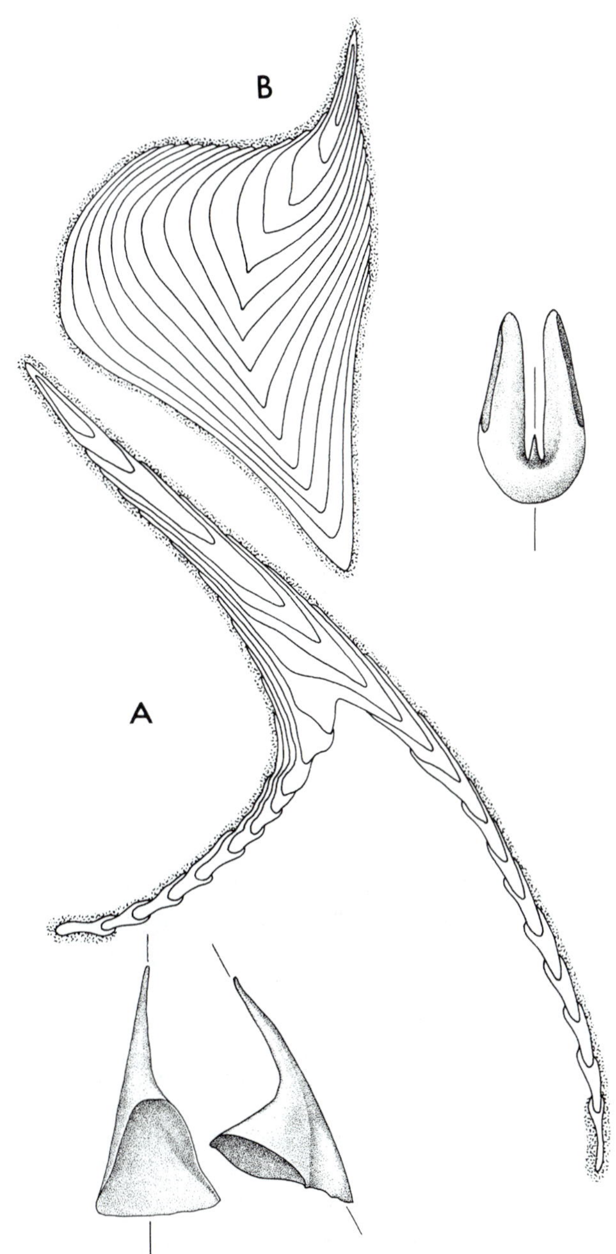

Fig. 6. Longitudinal sections through growth centers of earliest conodonts. A. *Furnishina furnishi* Müller, Upper Cambrian of Sweden. × 100. B. *Westergaardodina bicuspidata* Müller, Upper Cambrian of China. × 150. (After Müller and Nogami, 1971.)

The lower side of most conodonts is excavated. Two taxonomically significant structures can be distinguished:

(1) *The basal cavity*: an excavation which increases in size throughout growth of the specimen, its size being therefore dependent on the size of the specimen (Fig. 8).

(2) *The pit*: which formed when growth proceeded laterally from or above the initial point of secretion. The size does not increase throughout ontogenetic growth and ceases after a number of growth lamellae have been formed. Large specimens may thus have small pits (Fig. 9).

The ontogeny of individual conodonts can be studied in thin sections. Initial stages of all types correspond to the simple-cone type. In compound conodonts the various denticles present have developed from this simple structure through the thickening of the growth lamellae in well-defined directions of prevalent growth and by condensing them in intermediate directions. This simple manner of developing highly complicated shapes is called **anisometric growth.** Commonly, the direction of accelerated growth changes during ontogeny, and may lead to the development of peculiar shapes which are of taxonomic importance.

Structures within the denticles have been named **"white matter",** because in translucent fossils they appear white under reflected light (Fig. 10). In transmitted light, however, they appear as dark areas.

The upper surface of most conodonts is adorned with a sculpture of nodes or ridges. In addition, very small ridges may be present on the upper surface and are arranged in striae, polygons, etc. Because of their small size, the use of scanning electron microscope is necessary for study of these features which may prove useful in taxonomy (Fig. 11).

Conodont assemblages

Hinde (1879), while observing conodonts

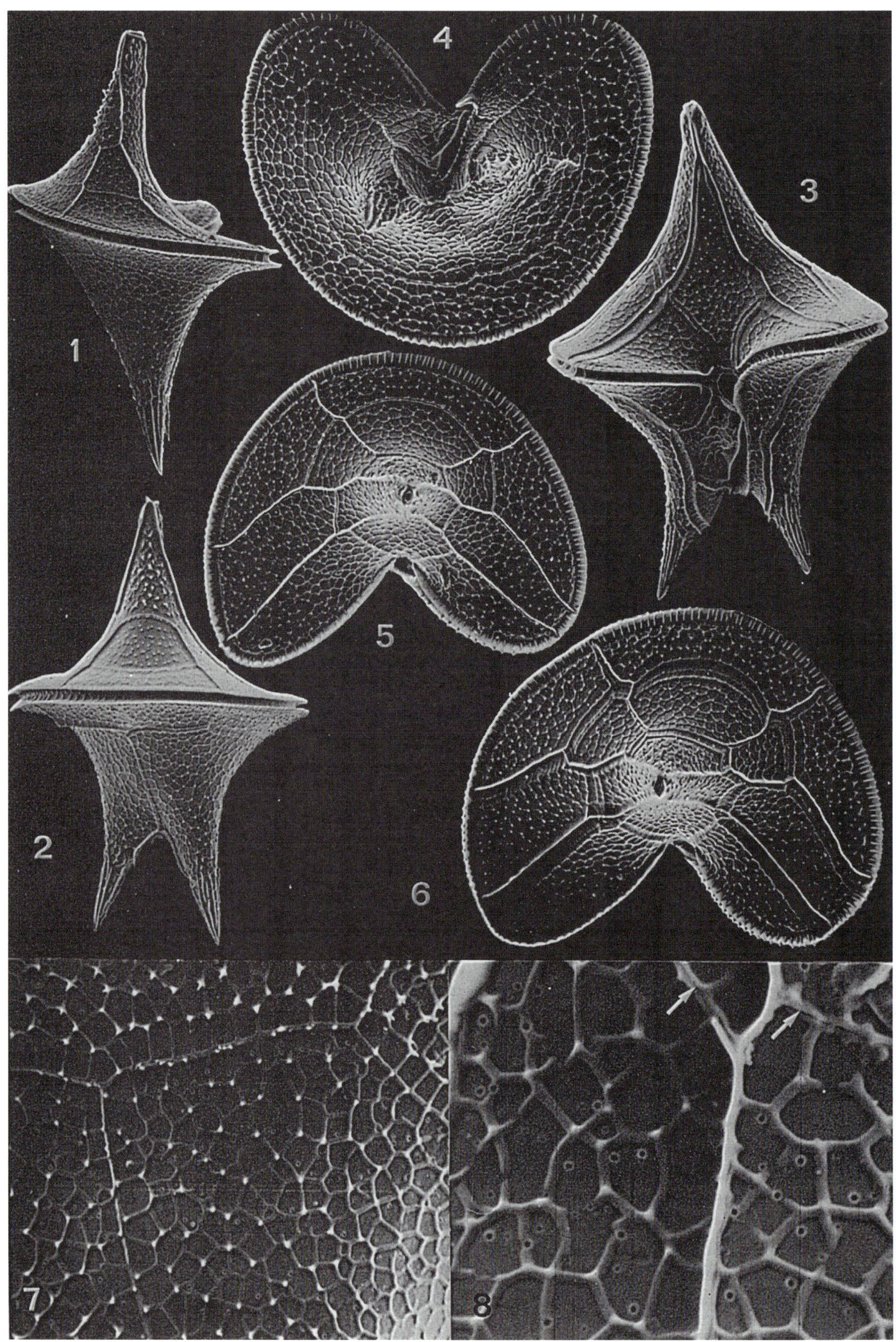

Fig. 5

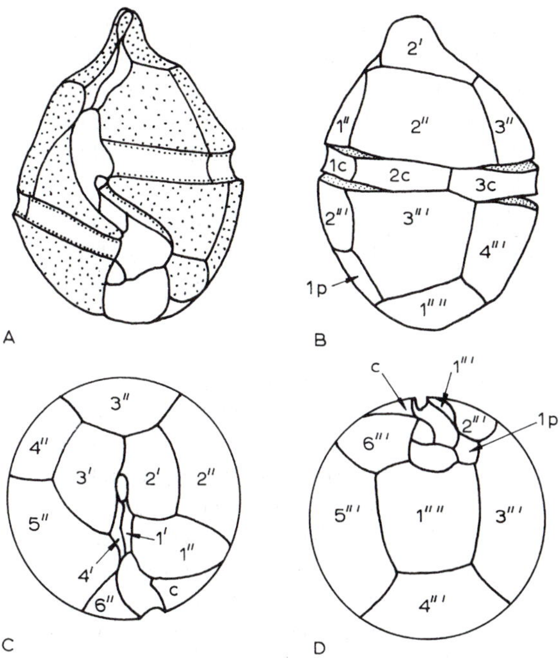

Fig. 6. Theca of *Gonyaulax spinifera* (Claparède and Lachmann) Diesing, showing tabulation. The tabulation may be expressed, thus, 4′, 6″, ?6c, 6″ ′, 1″ ″, 1p. *c* = cingular; *p* = posterior intercalary. A. Ventral view. B. Lateral view. C. Apical view. D. Antapical view. (From Wall and Dale, 1970.)

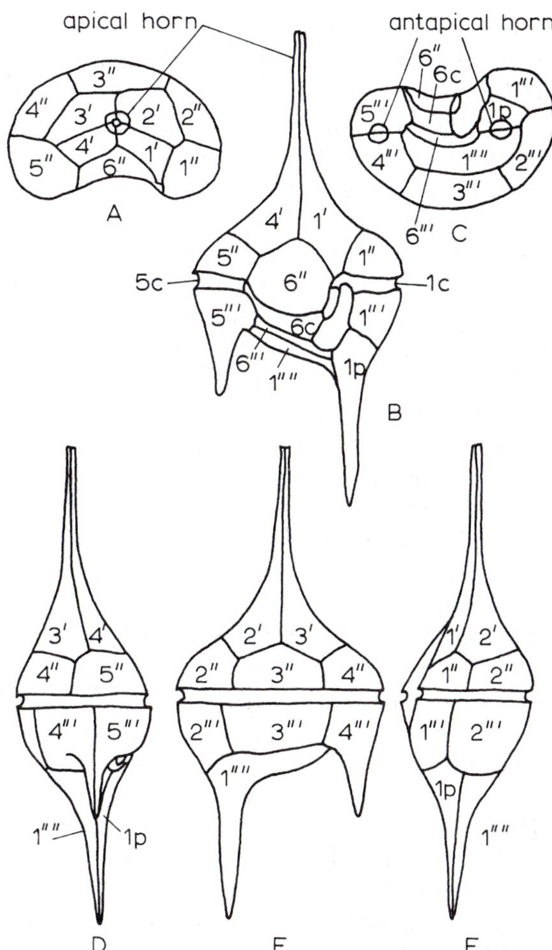

Fig. 7. *Ceratium* sp., showing tabulation of the theca or vegetative stage. The tabulation may be expressed, thus, 4′, 6″, 5–6c, 6″ ′, 1p, 1″ ″. A. Apical view. The circle denotes the apical horn. B. Ventral view. Note the offset sulcus. C. Antapical view. The circles denote the antapical horns. D. Right lateral view. E. Dorsal view. F. Left lateral view. (Figure courtesy W.R. Evitt.)

Ecology

The ecology of living dinoflagellates is very imperfectly known and we can only generalize about distribution patterns, with data based almost exclusively on the motile stage of oceanic species. Dinoflagellates are found in all environments from fresh water to open oceanic. The motile stages of autotrophic forms live in the photic zone. Species can be divided ecologically according to temperature, into **eurythermal** or temperature-tolerant species which are cosmopolitan, and **stenothermal** or temperature-sensitive species which usually are restricted to warmer waters. The distribution of the genus *Ceratium* is controlled in part by phosphate content of the sea water and ocean current systems.

The abundance of dinoflagellates fluctuates seasonally. Commonly, they multiply most

Fig. 5. SEM micrographs of the theca of the living species *Peridinium grande* (reproduced from Gocht and Netzel, 1974, by courtesy of the authors). 1. Young theca. × 290. 2. Theca with narrow dorsal intercalary bands. × 290. 3. Theca with broad ventral intercalary bands. × 300. 4. Theca with narrow antapical intercalary bands. × 380. 5. Theca with narrow apical intercalary bands. × 345. 6. Theca with broad apical intercalary bands. × 350. 7 and 8. Enlargement to show surface relief and pores. 7. Dorsal view, hypotheca of a young specimen, showing sutural ridges and reticulum. × 750. 8. Close-up of another specimen; sporadic separation of sutural ridges indicated by arrows. × 1900.

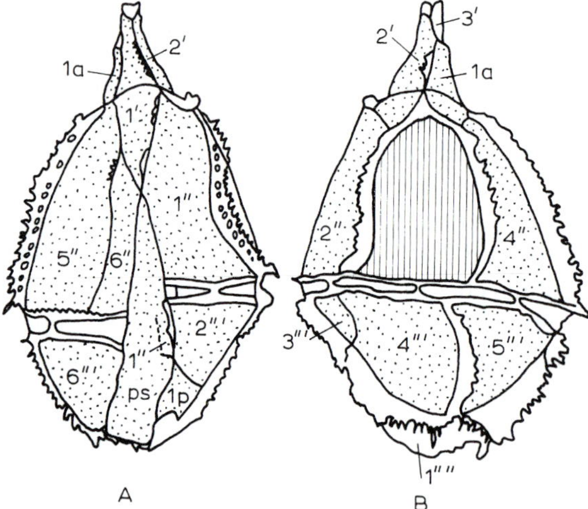

Fig. 9. Paratabulation of a proximate dinocyst, *Gonyaulacysta* cf. *Jurassica* (Deflandre) Norris and Sarjeant, from the late Jurassic of England. A. Ventral view. B. Dorsal view. *Ps* = parasulcus. The shaded area denotes the opening or archeopyle which has been formed from the loss of paraplate 3″. Note that Kofoid's alphanumeric system has been adopted for denoting the cyst paratabulation.

Chorate cysts possess a main body bearing strongly developed ornamentation or processes, as in *Hystrichosphaeridium* (Fig. 10).

A third major cyst type, the **cavate cyst** (Fig. 11), has inner and outer bodies that are only in limited contact, with intervening spaces between them. This group includes many cysts of the peridinioid type such as *Deflandrea* (see Figs. 25D and 26C). These cyst types are useful for morphologic distinction, but are only descriptive terms and should not be taken to indicate genetic affinities.

Shape and wall structure

Dinocysts are variable in shape. They may be spherical to ovoidal to ellipsoidal to elongate or peridinioid. Original flattening in a dorso-ventral plane is found in many genera, for example *Cyclonephelium*.

The wall of the dinocyst consists of one or more microscopic layers that form one or more bodies whose enclosure, one by another, isolates one or more cavities. By combining one of four prefixes designating spatial relationship with one of three suffixes, which refer either to the wall layer, the three-dimensional structure formed by that wall layer, or the enclosed cavity, twelve descriptive terms can be formed (Table III, Fig. 12). Thus, in a dinocyst with one recognizable wall layer, the **autophragm**, the body formed by this is the **autocyst** and the enclosed cavity is the **autocoel** (Fig. 12A). In bilayered forms there is one inner cavity, the **endocoel**, and one to several outer cavities or **pericoels** (Fig. 12B). The endocoel is completely closed off

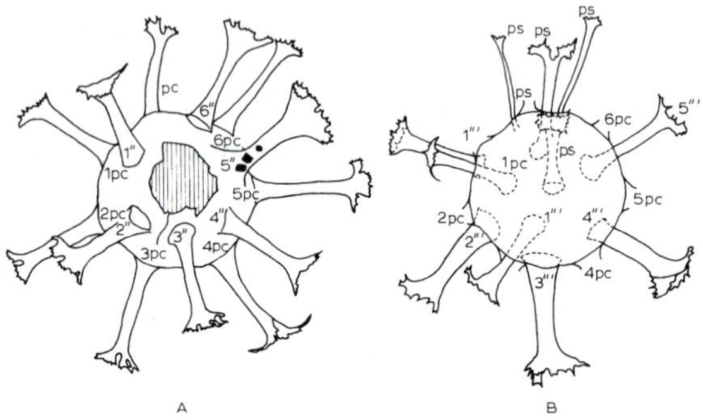

Fig. 10. Paratabulation of a chorate cyst, *Hystrichosphaeridium tubiferum* (Ehrenberg) Deflandre, as inferred from the number and position of the intratabular processes. A. Upper surface showing apical tetratabular archeopyle (shaded), precingular processes (1″ etc.) and paracingular processes (*pc*). B. Lower surface seen through the upper, showing postcingular processes (1″ ′ etc.), parasulcal processes (*ps*), single antapical process (1″ ″), and posterior intercalary processes (*p*). This is a chorate cyst. Early Tertiary, England.

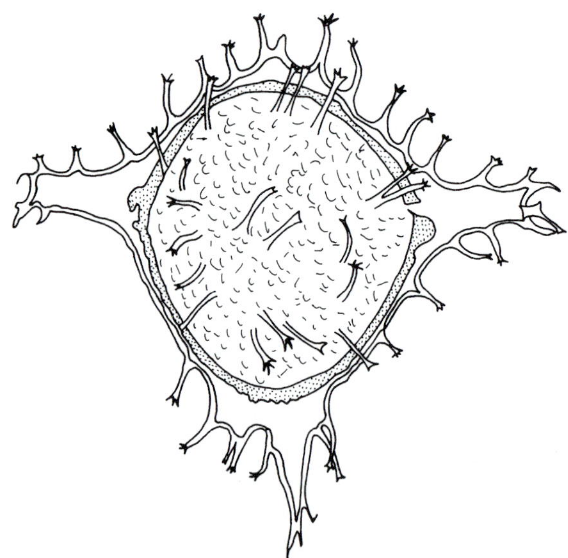

Fig. 11. *Wetzeliella lunaris* Gocht, a cavate cyst. Early Tertiary, England. × 480.

TABLE III

Wall terminology for dinocysts

	Meaning	Combination	Definition
Prefix			
auto-	single	autophragm autocyst autocoel	single wall single body single cavity
endo-	inner	endophragm endocyst endocoel	inner wall inner body inner cavity
meso-	middle	mesophragm mesocyst mesocoel	middle wall middle body middle cavity
peri-	outer	periphragm pericyst pericoel	outer wall outer body outer cavity
Suffix			
-phragm	wall	autophragm endophragm mesophragm periphragm	single wall inner wall middle wall outer wall
-cyst	body	autocyst endocyst mesocyst pericyst	single body inner body middle body outer body
-coel	cavity	autocoel endocoel mesocoel pericoel	single cavity inner cavity middle cavity outer cavity

from the outside if the archeopyle is not developed.

Paratabulation and ornamentation

In living species, a cyst often shows the same tabulation as the theca from which it was derived and if so, this is called reflected tabulation or paratabulation in the cyst to distinguish between cyst and theca. The alphanumeric system used to describe the thecal plates is, however, also used for the paratabulation in cysts. Similarly, the "plates" in cysts are called **paraplates** here and the principal paraplate series and intercalaries in cysts are shown in Fig. 13 and Table II. The cingular paraplates in cysts collectively form the **paracingulum** or **girdle**. The paracingulum divides the cyst into the **epicyst** and **hypocyst**. (These last two terms are the cyst equivalents of epitheca and hypotheca, respectively.)

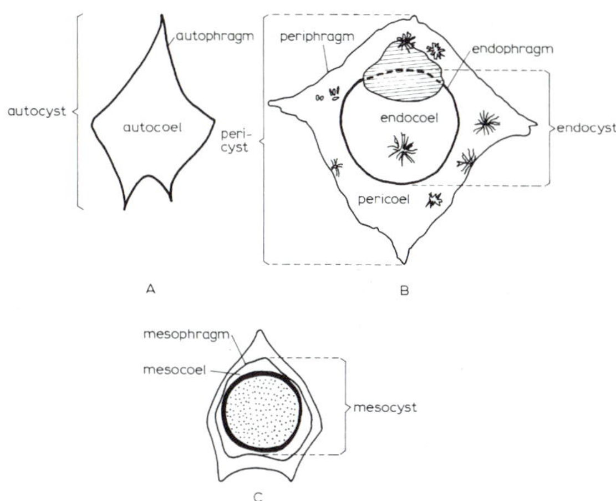

Fig 12. Body, wall and cavity terminology in dinocysts: A. *Lejeunia* sp. B. *Rhombodinium glabrum* (Cookson) Vozzhennikova. C. *Deflandrea* sp.

Some cysts have a smooth surface; others possess ornamentation which can be minor elements such as granules or major elements like horns, septa or processes. A dinocyst never has more horns than its corresponding thecate stage and in both the maximum is five. Horns may be apical, antapical or lateral only (Fig. 14) and are useful criteria for the orientation of the dinocyst.

Processes are essentially columnar or spine-like (Fig. 15), whereas septa are membraneous, linear projections arising perpendicularly from the outer wall layer (see Figs. 9 and 18A). Processes may be located on or within paraplate boundaries. **Process complexes** are processes on individual paraplates that are united proximally, distally, or along their length (Fig. 16).

All the processes of a dinocyst may be similar (see Fig. 19B) or they may be differentiated. Process differentiation is not haphazard and a particular paraplate series often is characterized by a diagnostic type of process. In other taxa some of the paraplates, commonly the cingular, are devoid of processes (Fig. 15B). The preferred orientation of septa and processes often permits one to determine paratabulation in dinocysts. Process arrangements, shapes and terminations are important morphologic features in generic and specific classification.

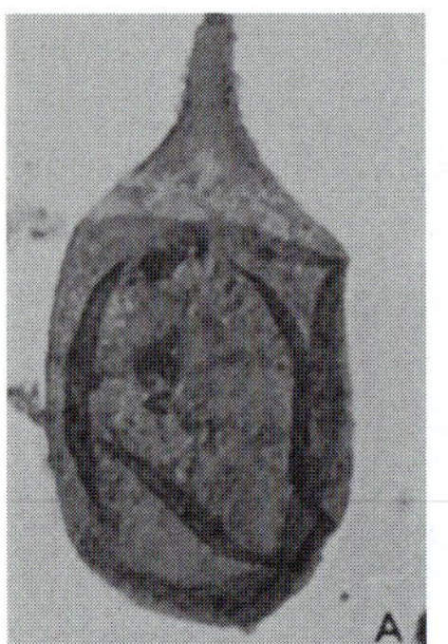

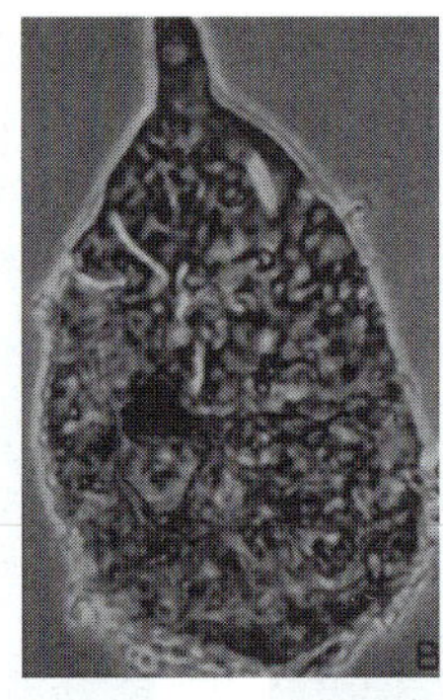

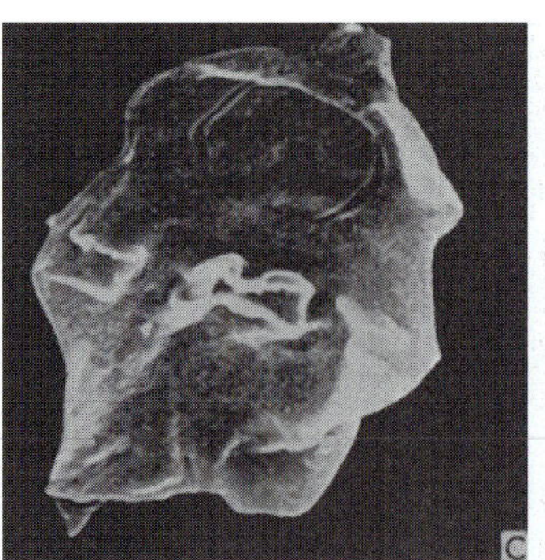

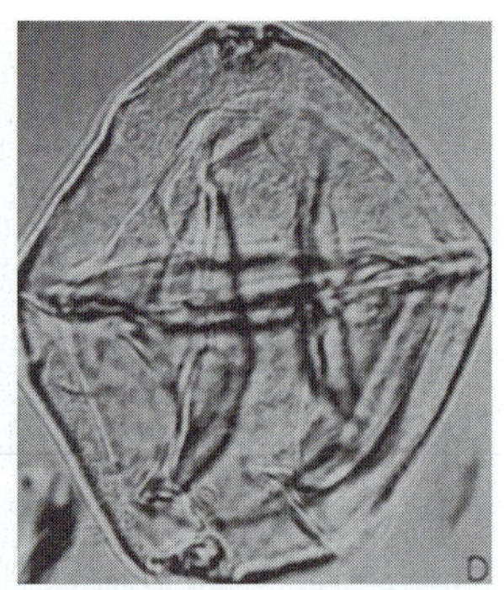

Fig. 25. Late Jurassic and Cretaceous species of the peridinioid lineage. A. *Pareodinia ceratophora* Deflandre, late Jurassic. × 650. B. *Imbatodinium* sp., early Cretaceous. This differs from *Pareodinia* only in possessing processes. × 640. C. *Australiella victoriensis* (Cookson and Manum) Lentin and Williams, late Cretaceous. Archeopyle intercalary with operculum remaining attached along posterior margin. × 640. D. *Deflandrea cretacea* Cookson, late Cretaceous. The operculum of the intercalary archeopyle remains attached along its posterior margin. × 805.

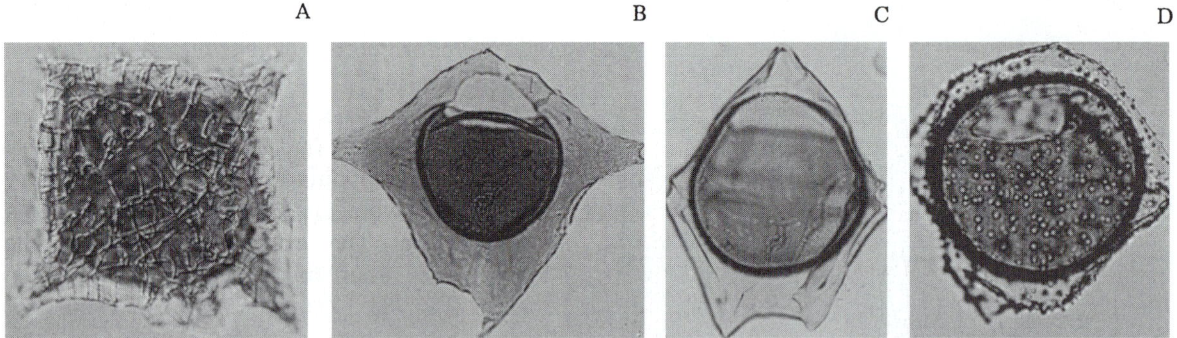

Fig. 26. Paleogene examples of the peridinioid lineage. A. *Wetzeliella tenuivirgula* subsp. *Crassoramosa* (Williams and Downie) Lentin and Williams. The simulate process complexes are particularly noticeable on the epicyst. × 940. B. *Rhombodinium glabrum* (Cookson) Vozzhennikova. Dorsal view, archeopyle intercalary. × 300. C. *Deflandrea phosphoritica* Eisenack. Archeopyle intercalary. × 300. D. *Wetzeliella condylos* Williams and Downie. Dorsal view, archeopyle intercalary. × 350.

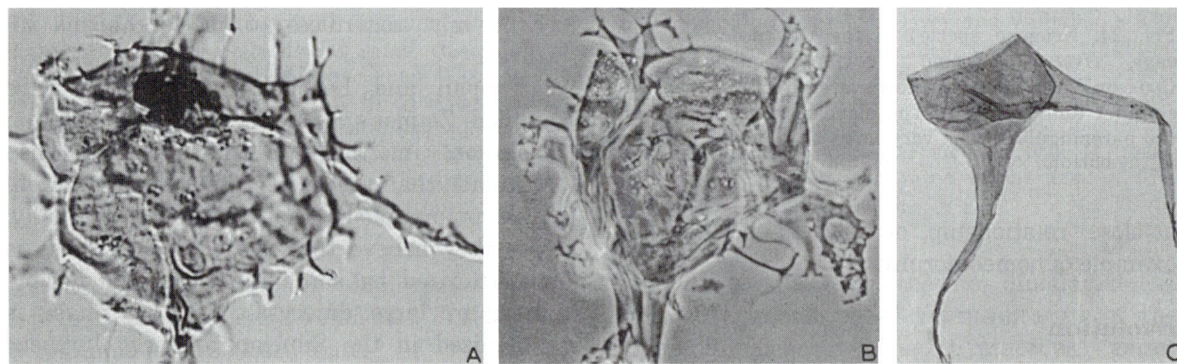

Fig. 27. Cretaceous examples of the ceratioid lineage. A. *Phoberocysta neocomica* (Gocht) Millioud, early Cretaceous. Dorsal view with apical archeopyle visible. × 670. B. *Xenascus ceratioides* (Deflandre) Lentin and Williams, late Cretaceous. Archeopyle apical. × 370. C. *Odontochitina operculata* (O. Wetzel) Deflandre and Cookson, late Cretaceous. Archeopyle apical. × 200.

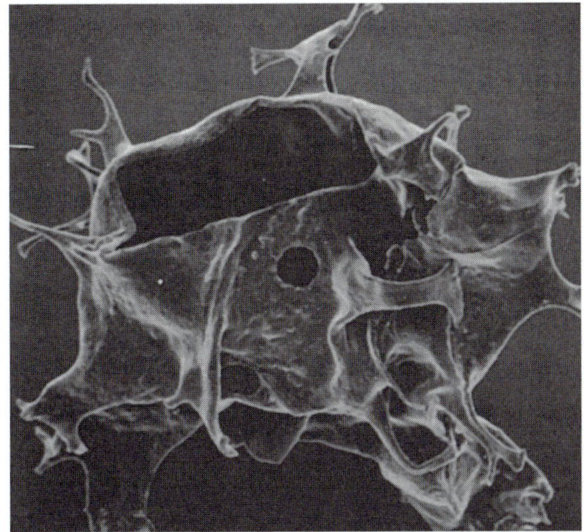

Fig. 28. Electron micrograph of the late Cretaceous species of the ceratioid lineage, *Xenascus ceratioides* (Deflandre) Lentin and Williams, showing the apical archeopyle towards the top. (cf. With Fig. 28B). × 1000.

second cycle, characterized by a separate evolutionary group commenced its explosive phase in the early Cretaceous and continued for 25 m.y. before it reached its climax. The late Cretaceous marked a decline phase. A third explosive phase commencing in the late Paleocene and extending into the early Eocene was followed by a decline phase, which extends to the Recent (Fig. 35). If we assume a cycle occupies 50 m.y. it is possible that Recent dinoflagellates are either experiencing or approaching the evolutionary phase of a fourth cycle.

The death phase of one cycle tends to be succeeded by the explosive phase of another cycle and these explosive cycles often seem to be related to maximum development of epicontinental seas. Other major environmental changes may also act as triggering mechanisms.

From the data generated by Sarjeant

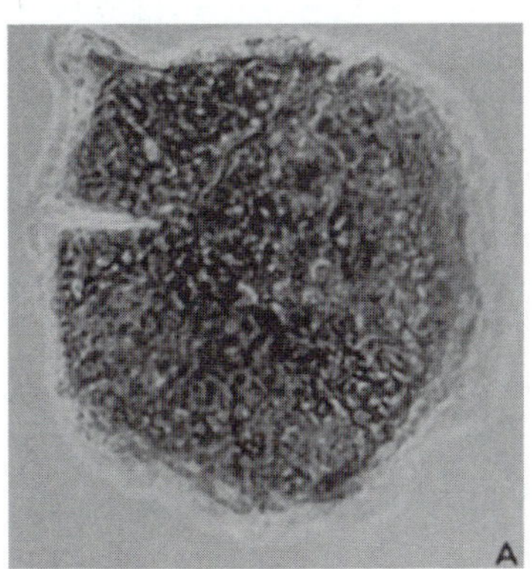

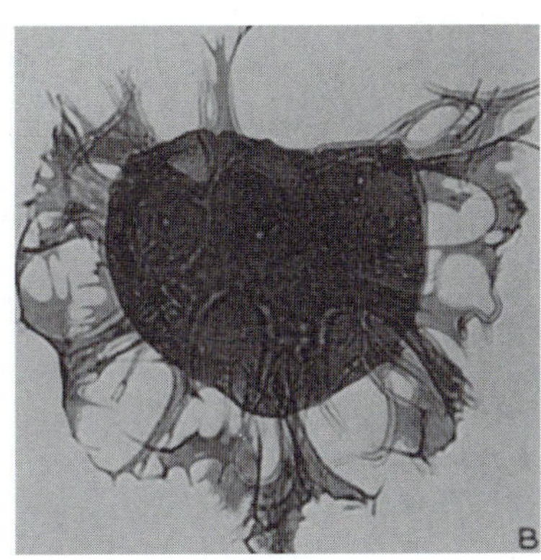

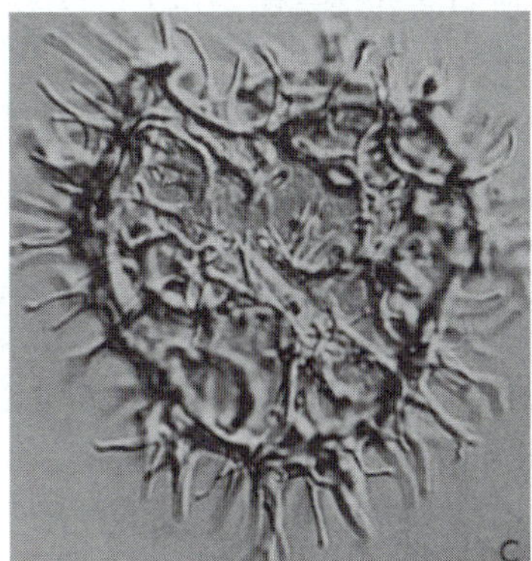

Fig. 29. Examples of the *Cyclonephelium* lineage. A. *Cyclonephelium vannophorum* Davey, late Cretaceous. The operculum of the apical archeopyle is slightly displaced. × 600. B. *Cyclonephelium ordinatum* Williams and Downie, Paleogene. The apical archeopyle permits orientation of the specimen. × 485. C. *Areoligera senonensis* Lejeune-Carpentier, Paleogene. The annulate process complexes delineate the paratabulation. × 485.

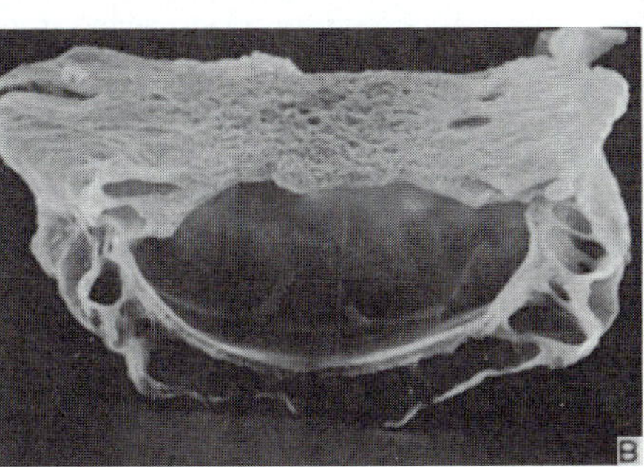

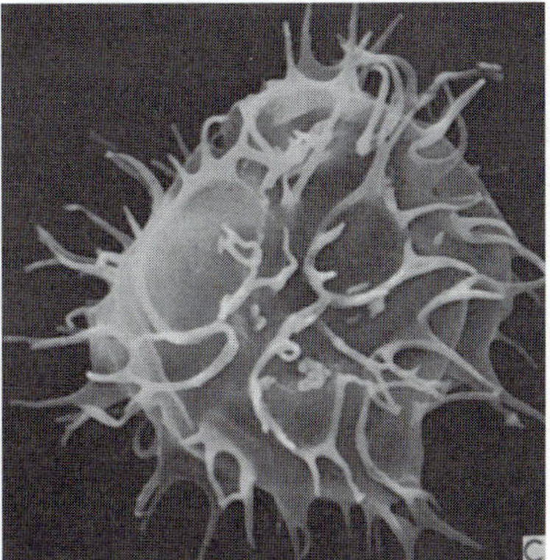

Fig. 30. Examples of the *Cyclonephelium* lineage. A. *Senoniasphaera protrusa* Clarke and Verdier, late Cretaceous. Dorsal view with apical archeopyle at top; the longer antapical horn is at the left. × 800. B. *Chiropteridium aspinatum* (Gerlach) Brosius, Paleogene. Apical view looking into the apical archeopyle. × 800. C. *Areoligera senonensis* Lejeune-Carpentier, Paelogene. Dorsal view showing annulate process complexes. × 800.

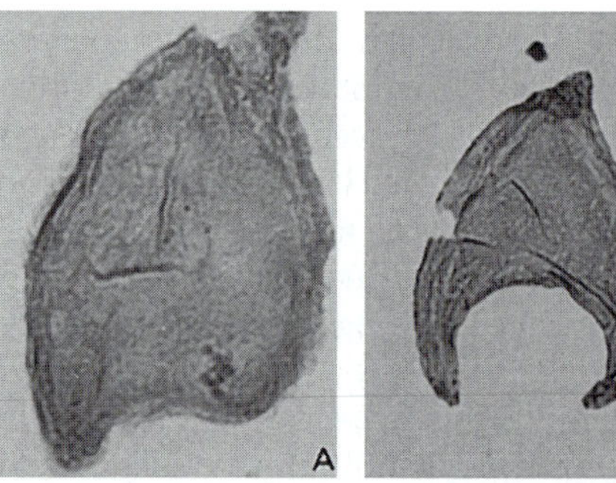

Fig. 31. Examples of *Nannoceratopsis* lineage. A. *Nannoceratopsis gracilis* Alberti, early Jurassic. The break in slope denotes the paracingulum. × 600. B. *Nannoceratopsis pellucida* Deflandre, late Jurassic. The two long curved horns are antapical. × 325.

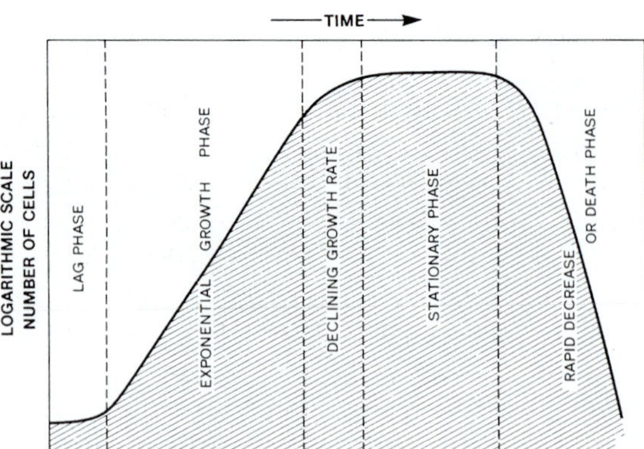

Fig. 34. Growth phases of algal protists in a limited culture, from inoculation to death. (From Tappan and Loeblich, 1970.)

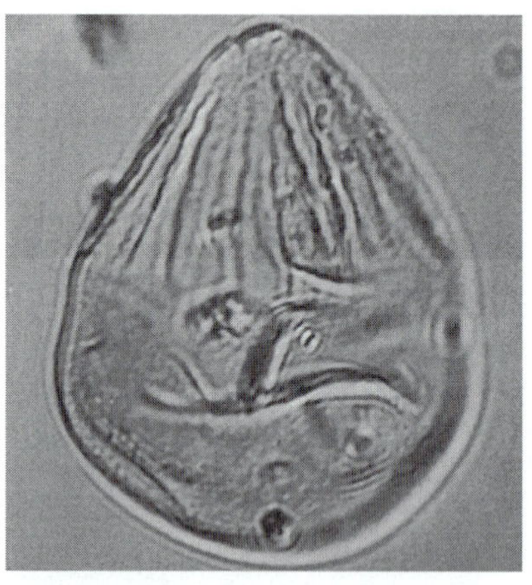

Fig. 32. Example of the *Dinogymnium* lineage, *Dinogymnium euclaensis* Cookson and Eisenack, late Cretaceous. × 1,350.

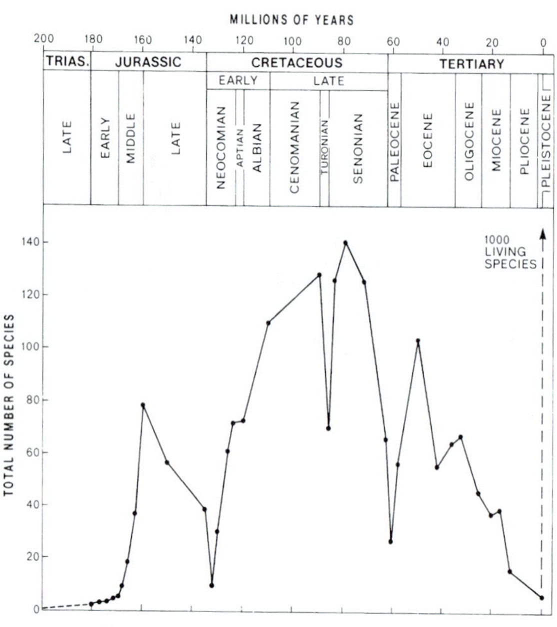

Fig. 35. Variations in number of dinocyst species in the Mesozoic–Cenozoic. Points denote end of period for which species were totaled. (From Tappan and Loeblich, 1970.)

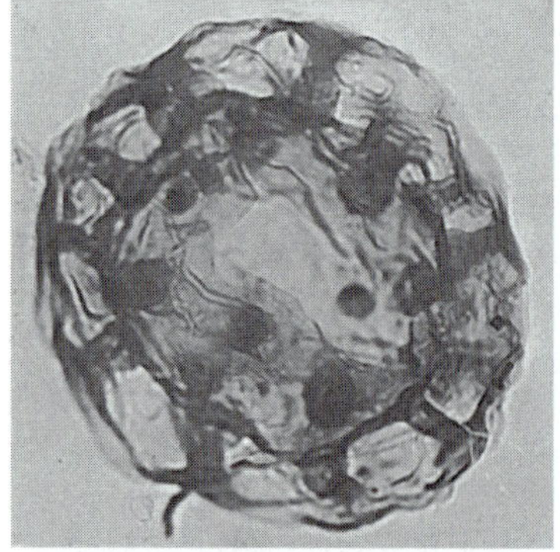

Fig. 33. Example of the *Tuberculodinium* lineage, *Tuberculodinium vancampoae* (Rossignol) Wall, Neogene. Archeopyle hypocystal. × 500.

(1967), Tappan and Loeblich estimate that a new fossil dinocyst genus appeared every 1.4 m.y. and that the average duration of a genus was 42.4 m.y. Including the number of modern genera lowers the post-Paleozoic evolution rate to one genus every 460,000 years. Such a rapid turnover suggests that they are ideally suited for biostratigraphic zonation.

Phyletic trends

Primitive features of paratabulation appear to be observable in some lineages. In the peridinioid lineage, genera with up to six anterior intercalaries, as *Pareodinia* (see Fig. 25A) are present in the early Jurassic and this apparently is a primitive feature. Other features of paratabulation are remarkably constant. The tendency for the first post-cingular to be reduced and incorporated into the sulcal region is found in the *Gonyaulacysta* lineage throughout its known stratigraphic range and is possibly also a feature of *Ceratium*. Eaton (1971) has recognized an evolutionary lineage in some dinocyst species from the Eocene of southeast England (Fig. 36). Such lineages enhance the value of dinocysts in biostratigraphic studies.

The recognition of evolutionary trends in archeopyle types also must be largely conjectural, although Lister (1970b) has recognized a sequential development. In the most primitive stage, the excystment aperture is defined by a suture that has no surface manifestation before dehiscence. Thus, the transapical suture present in *Peridinium limbatum* (see Fig. 17) is a primitive feature and perhaps this species is a long-ranging stock. A more advanced stage is represented by species with an archeopyle margin that is not yet polygonal. Species with a polygonal archeopyle margin represent the most advanced stage in the development of the archeopyle. This last category would presumably include apical archeopyles although apical archeopyles are not known in present-day *Gonyaulax* species.

Biogeography and paleoecology

The distribution pattern of only one genus, *Ceratium*, is known in any detail. Unfortu-

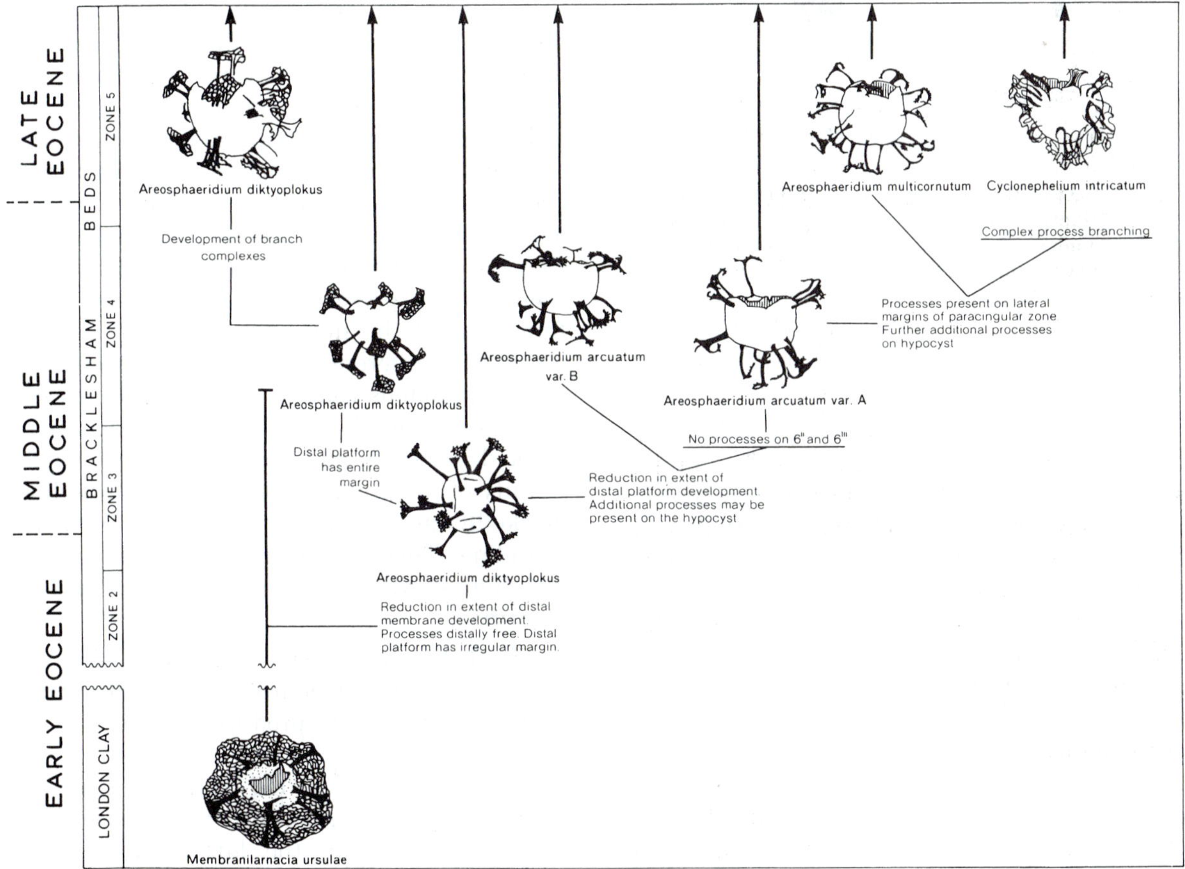

Fig. 36. The postulated evolution of the *Membranilarnacia ursulae* lineage in the Eocene of southeast England. (From Eaton, 1971.)

Spores

Spores produced by "lower plants", such groups as algae, fungi, mosses, and ferns, are some of the earliest preserved remains of plants (Fig. 2). The Silurian marks the appearance of the first accepted spores with triradiate sutures. Spore production in primitive vascular plants, such as the club moss, may be homosporous, that is, a single type of spore is formed; or heterosporous. Heterosporous plants, those that produce male microspores and female megaspores, are known since the late Devonian. (In a strict sense pollen grains are microspores and are analogous to the microspores of ferns.) In homosporous plants the spore germinates and develops into either bisexual gametophytes or into two plants one of which produces male gametophytes and one female gametophytes.

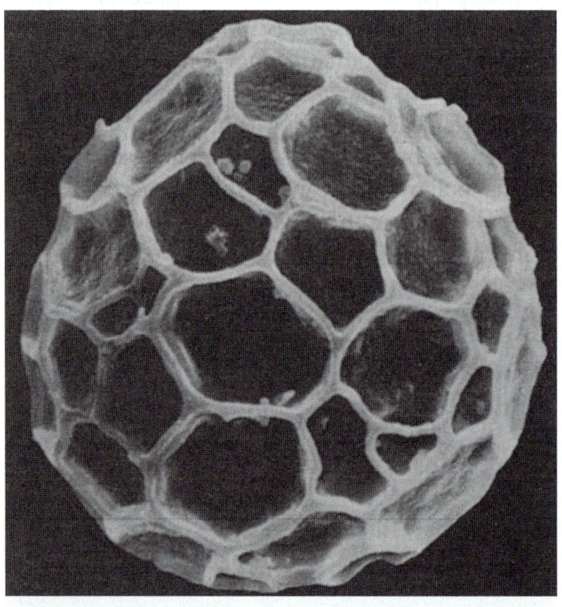

Fig. 2. A monad apore, *Lycopodium gayanum*. × 1500.

CHARACTERISTICS

Distinguishing criteria for pollen

The reader should consult one or more of the texts listed in "Suggestions for further reading" for a more extensive account of pollen and spore morphology and taxonomy. We will touch upon these aspects only briefly.

Pollen grains (and spores) are distinguished by the aspects of morphology — size, shape, apertures, surface sculpture — and wall structure.

Size

The size of the majority of pollen grains is between 20 and 80 μm, with rare forms less than 10 μm to more than 200 μm. Approximate size is relatively constant and useful in identification; however, it should be noted that size variations are caused by a number of factors, including preparatory technique. For example, mounting in glycerine gelatin tends to cause swelling of the pollen grains; the same grains embedded in silicone oil are about 30% smaller.

Shape

A wide variety of radially symmetrical shapes exist, of which variations of the rotational ellipsoid are more common. The relation between the polar and equatorial dimensions forms the basis for definition of commonly used shape classes, some of which are illustrated in Fig. 3. Variations of the ellipsoid form are frequently encountered.

Bilaterally symmetrical convex pollen grains are produced principally by two groups, the monocotyledons (lilies) and the gymnosperms. The latter produce vesiculate grains, those with sacci (bladders or wings) (Fig. 3F).

Apertures

The number and type of apertures (openings or thinning of part of the exine) form the primary basis for pollen differentiation. Generally, isodiametric apertures are called **pores**, and elongate or furrow-like apertures are called **colpi** (Fig. 4). Pollen grains may have no apertures (**inaperturate**), single apertures (**monoporate** or **monocolpate**), multiple pores (**diporate, triporate, stephanoporate**, or **periporate**), multiple colpi (**dicolpate, tricolpate, stephanocolpate, heterocolpate**, or **syncolpate**), or multiple pores and colpi (**dicolporate,tricolporate, stephanocolporate, pericolporate**).

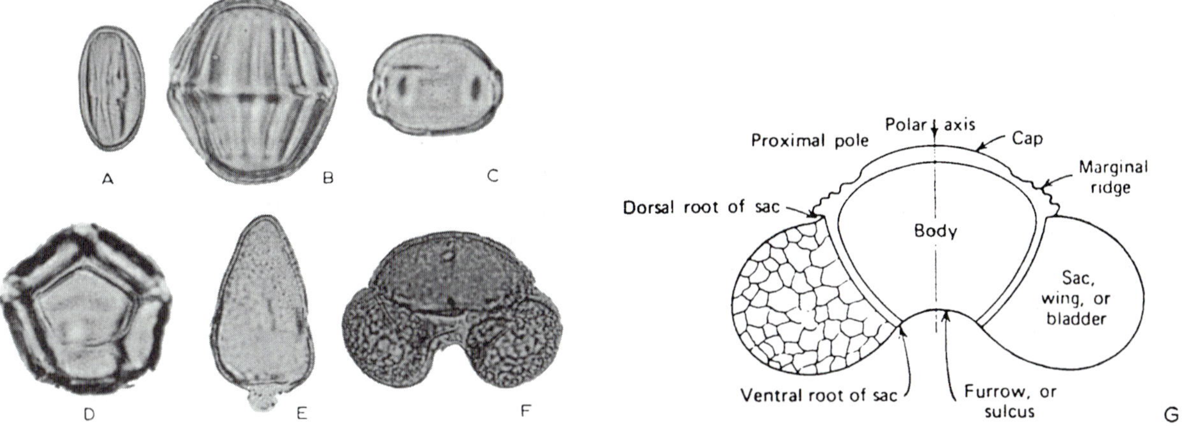

Fig. 3. Some common shapes of pollen grains. A. Prolate. B. Subspheroidal. C. Oblate. D. Pentagonal. E. Irregular. F. Vesiculate. G. Descriptive terms. (All figures approximately × 330, except D: × 890.)

Sculpture

Perhaps of greatest diagnostic value in the identification of pollen grains are surface sculpture and wall structures. Some twelve kinds of sculpture are recognized: these include types that are smooth, pitted, grooved, or ones that exhibit more or less isodiametric or elongate elements (see Fig. 4).

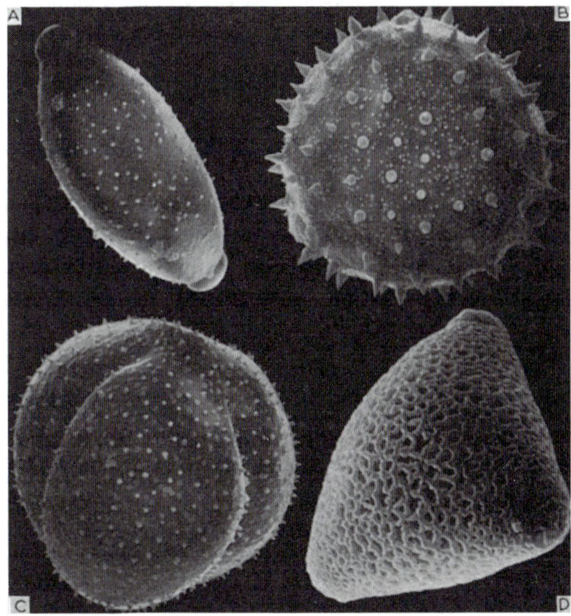

Fig. 4. Scanning micrographs showing some apertural types and sculpture on pollen grains. A. Diporate, microechinate grain of *Embothrium coccinineum*. × 1400. B. Tricolporate, echinate grain of *Corynabutilon* vitifolium. × 750. C. Triporate, foveolate–echinate grain of *Cereus chiloensis*, × 750. D. Triporate, echinate heterobrochate grain of *Lomatia ferruginea*. × 1500.

Wall structure

Sculptural elements are developed in the outer structural layer of the pollen grain, the **exine**, which is frequently divided into an inner homogeneous layer, the **endexine**, and an outer layer, the **ektexine** (Fig. 5). The exine is composed of a complex of minerals and organic compounds which include **sporopollenin**, a "natural plastic" which is highly resistant to degradation. The **intine** or interior wall of the pollen grain which encloses the protoplasm or living nucleus is largely composed of cellulose and other substances which are readily destroyed and thus are rarely fossilized.

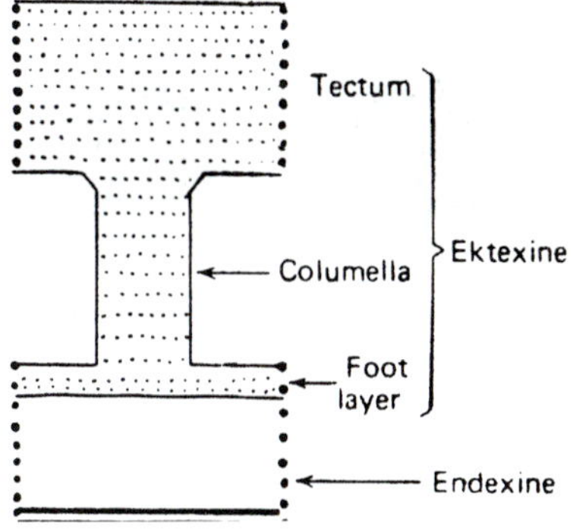

Fig. 5. Details of the pollen wall structure.

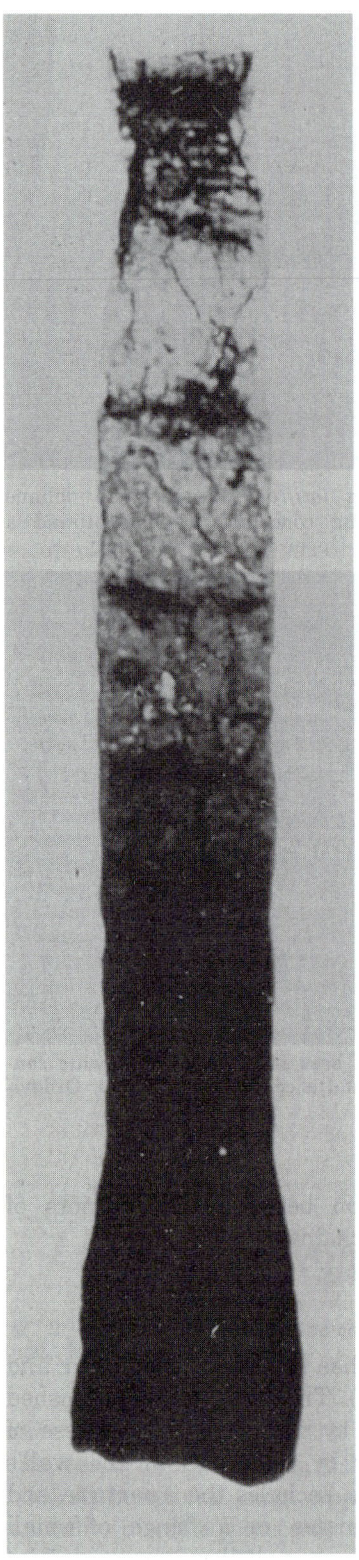

Fig. 9. *Conochitina parviventer* Jenkins, showing the rings or discs that make up the prosome, and the dark operculum sealing the aperture, in lateral view. × 450. Ordovician, Britain.

it is made up of a series of rings or discs and terminates orally in a disc-like **operculum** which may have sealed the contents of the test from the outside (Fig. 9). In some genera, however, particularly those without necks,

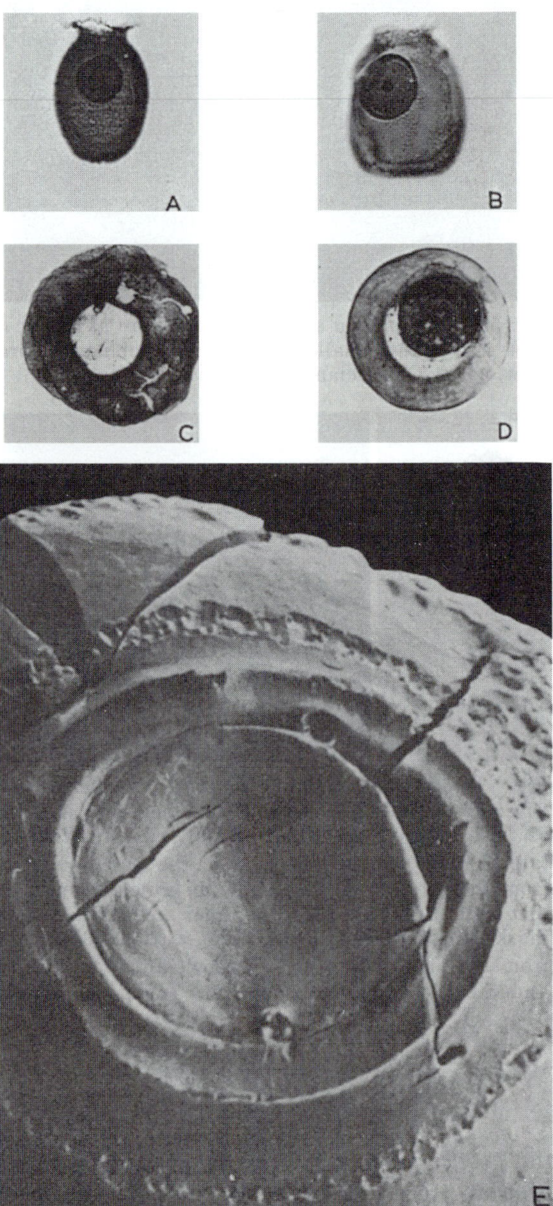

Fig. 10. Chitinozoan operculum. A, B. Two examples of *Desmochitina minor* Eisenack, in lateral view showing the operculum fallen into the chamber, × 250. C, D. Lenticular chitinozoans in oral view, showing (C) open aperture after the operculum has been lost, and (D) operculum lying partly acorss the aperture, × 200. E. Lenticular chitinozoan in oblique oral view showing operculum firmly in place. × 600. A, B and D from the Upper Ordovician of Oklahoma; C from the Ordovician of Britain; and E from the Lower Devonian of Pennsylvania.

the prosome lacks the rings or discs and consists solely of an operculum (Fig. 10). Sometimes an operculum shows a central pore and concentric ribbing that reflect the structure of the base to which it was once attached.

Test wall

In well-preserved material the wall of the chitinozoan test is translucent and amber-colored. Carbonized tests recovered from rocks altered by metamorphism are opaque, black and brittle.

Chitinozoan species with one, two and three wall layers are known (Laufeld, 1974) although in most species the wall is made up of two layers. The outer surface of the wall may be smooth or ornamented (Fig. 11).

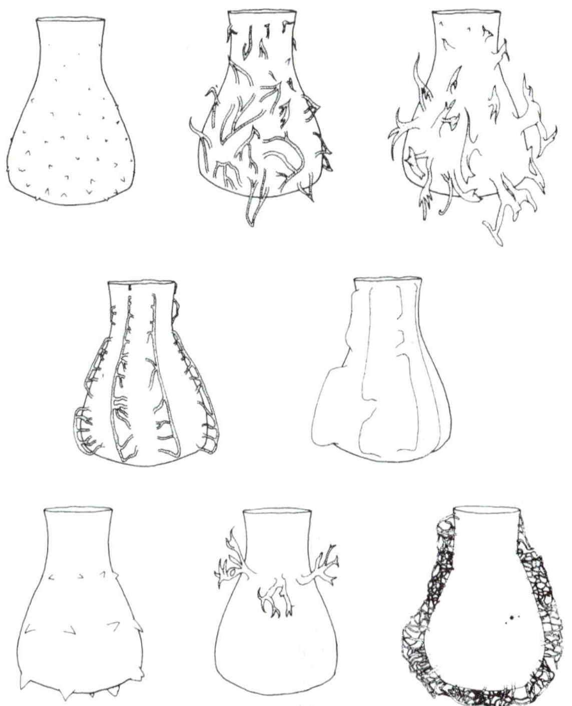

Fig. 11. Types of ornament (stippled) in lateral view. × 125.

Ornament generally is less strongly developed on the neck than on the chamber but there are important exceptions. Ornamental processes often stand in rows parallel with the length of the test (see Figs. 22, 28). Small ornamental processes are solid or have "spongy" interiors, whereas large processes are largely hollow. Like appendices, large ornamental processes do not open into the body cavity, and the surface of the test lying within the base of a large process does not differ from the adjacent surface outside the base of the process. These observations have led Laufeld (1974) to conclude that the test wall was formed before the appendices or spines were secreted on it. Processes may have served to control buoyancy or as a means of attachment.

OCCURRENCE OF CHITINOZOAN TESTS

Most chitinozoans occur in the rock as single tests (Fig. 12A). Frequently, however, two or more tests are found joined together in linear chains (Fig. 12B). Such chains are known in all the common genera, though in some they may be rare. It is believed that the capacity to form chains is a fundamental chitinozoan characteristic, although most species have not yet been found preserved in this condition. In some species chains are very long and form spiral coils (Fig. 12C). Occasionally, the whorls of a spiral chain comprising several hundred tests have been found packed tightly together and enclosed in an organic pellicle called a **cocoon** (Fig. 12D).

Collection of samples

The concentration of chitinozoans in sediments generally is less than 20 tests per gram of rock and may be less than one test per gram. The greatest recorded concentration is 984 tests per gram in Silurian rocks on Gotland. Comparable figures for other organic-walled microfossils, such as acritarchs and spores, could be in the thousands per gram. Consequently, rock samples collected specifically for chitinozoan studies should be about 500 g.

Preparation of samples

Chitinozoans are prepared for microscopy In much the same way as other organic-walled microfossils. Care must be taken not to damage the relatively large and brittle tests which often bear delicate ornamentation. The

CYTOPLASM: all the protoplasm of a cell apart from the nucleus.

DENDRITIC: branched, tree-like.
DENTICLE: spine-like, needle-like, or sawtooth-like structure, similar to cusp but commonly smaller.
DEXTRAL: right-handed; term originally applied to any shell with aperture on observer's right when shell apex is directed upward, or with apparent clockwise coiling when viewed from above apex.
DIAGENESIS: all chemical, physical, and biological modifications undergone by a sediment after its initial deposition.
DIAGONAL SECTION: slice cutting axis of coiling obliquely.
DIATOMITE: a rock with diatoms as the most conspicuous component, generally thought of as a soft, fine-grained, porous, low-density material which is white or very light in color.
DICTYOSOME: a unit of Golgi apparatus several of which occur as discrete bodies in cells of invertebrates and plants.
DIMORPHISM: refers to two morphologically distinct phases or forms of a species.
DINOCYST: the resting cyst of a dinoflagellate. All fossil dinoflagellates appear to be dinocysts.
DIPLOID STAGE: having the chromosomes in pairs, the members of each pair being homologous, so that twice the *haploid* number is present.
DIPORATE: possessing two pores and no colpi.
DISTAL: that part of a spore or pollen grain that faces to the outside when a member of a tetrad.
DOLON: part of velar or histial structure which is modified in heteromorphs in comparison with that of adult tecnomorphs.
DOMICILIUM: part of the carapace exclusive of space enclosed by velar structures. In wider sense the part of the carapace enclosing the body of the animal, but excluding any kind of sculptural extensions of the valve.
DORSAL: the surface directly opposite to that containing the sulcus in a motile dinoflagellate, and the parasulcus in a dinocyst.
DORSAL: opposite to ventral side; spiral side of trochoid test.
DORSAL MUSCLE SCAR FIELD: imprints of muscles below the dorsum, including scars of different origin.
DUPLICATURE: in zoological works, synonymous with inner lamella. In paleontological literature, terms of recalcified part of inner lamella.

ECDYSIS: molting of body cover in arthropods.
ECHINATE: sculpture type of pointed elements at least 1 μm in height.
ECHINUS (pl. echini): component element of echinate sculpture.
ECTODERM: the outer layer present in acritarchs with bilayered walls.
ECTOPHRAGM: the outermost layer present in dinocysts with three walls in which the other two walls are the endophragm and periphragm, respectively.
EKTEXINE: usually the outermost layer of the exine when both the ektexine and endexine, or innermost layer, are represented; see perine.
ELLIPSOIDAL: shape like an ellipsoid; a solid the plane sections of which are largely ellipses but also circles.
ENCYSTMENT: the enclosure of nucleus and cytoplasm within a resistant wall commonly as a response to changing environmental conditions or as a preliminary to cell division. The cyst is formed within the amphiesma, but may grow to greater dimensions after the latter has decayed.
ENDEXINE: innermost layer of the exine.
ENDOCOEL: the cavity enclosed by the endophragm in a dinocyst.
ENDOCYST: the body formed by the endophragm in a dinocyst.
ENDODERM: the innermost wall in acritarchs with two or more walls.
ENDOPHRAGM: the innermost wall in dinocysts with two or more walls.
ENDOPLASM: central part of cytoplasmic mass, commonly granulated.
ENTOSOLENIAN: having internal tube-like apertural extension.
EPICYST: the portion of a dinocyst corresponding to the epitheca of a motile dinoflagellate. In paratabulate dinocysts this is the area anterior to the paracingulum and consisting of the apical, anterior intercalary (if present) and precingular paraplates.
EPICYSTAL ARCHEOPYLE: an archeopyle involving the loss of the epicyst and in which the archeopyle suture runs immediately anterior to the paracingulum.
EPISOME: in naked or unarmored motile dinoflagellates, the portion of the amphiesma which is anterior to the cingulum.
EPITHECA: also called the epivalve. The larger of the two valves is the frustule, it fits over the smaller valve. In a thecate motile dinoflagellate, the portion of the theca which is anterior to the cingulum. It includes the apical, anterior intercalary (if present) and precingular plates.
EPITYCHE: an outfolded flap of the vesicle wall in acritarchs.
EQUATOR: imaginary line, situated approximately midway between the polar extremities, that encircles symmetrical spores and pollen.
EQUATORIAL AXIS: imaginary line, perpendicular to the polar axis, that extends to the equator; in radiosymmetric pollen and spores the axis is more or less equidistant, whereas in bilateral types a short axis is at right angles to a long axis.
EUCARYOTES: morphologically advanced organisms characterized by having the cellular organelles, including the nucleus, bounded by membranes.
EUPHOTIC ZONE: zone near surface of sea (roughly upper 100 m) into which sufficient light penetrates for photosynthesis.
EUTROPHIC: literally "well nourished". A condition where there are enough nutrients in the aquatic environment to support the population.
EVOLUTE: tending to uncoil; chambers non-embracing.
EXCYSTMENT: in dinoflagellates, the process of cyst abandonment, which may be as a response to improving environmental conditions or after completion of cell division. Excystment occurs through an archeopyle or other preformed opening; the cell initially has a soft flexible wall lacking any differentiation of plates, etc., but quickly develops a more rigid amphiesma.
EXCYSTMENT APERTURE: the aperture through which the dinoflagellate escapes on excystment. Partly synonymous with archeopyle.
EXINE: outermost resistant layer of the pollen or spore wall.
EXOPLASM: the outer zone of radiolarian cytoplasm, outside the central capsule.
EXTRACARDINAL MARGIN: part of the valve margin outside the cardinal (hinge) margin. Syn.: free margin.
EXTRAUMBILICAL APERTURE: opening in final chamber of test not connecting with umbilicus, commonly sutural midway between umbilicus and periphery.

FACIES: the general aspect, nature, or appearance of a sedimentary rock.
FEET: in nassellarian radiolarians, terminal spines, few in number and usually relatively robust, directed away from the apex, often homologs of dorsal and primary lateral spines.
FILAMENT: a slender row or line of cells.
FILOPODIA: thin, delicate, needle-shaped pseudopodia.
FISTULOSE: having tubular irregular growth.
FLAGELLATED: bearing a flagellum.

FLAGELLUM: a thin whip-like structure capable of undulatory movement.
FLANGE: ridge along the valve margin of some podocopids made by projection of outer lamella, on the outer side of the selvage.
FORAMEN (pl. foramina): opening between chambers located at base of septa or areal in position.
FREE MARGIN: see extracardinal margin.
FREE OPERCULUM: an operculum which is completely detached from the rest of the dinocyst.
FRONTAL MUSCLE SCARS: scars situated in front of the adductor muscle scars and above mandibular scars. In podocopids hitherto studied, they correspond to imprints of a mandibular muscle and muscle attaching itself to chitinous endoskeleton. In earlier literature, they were erroneously interpreted as antennal muscle scars.
FRUSTULE: the individual diatom cell consisting of the epivalve, the hypovalve and a connecting band.

GAMETANGIUM: an organ producing gametes (which form the zygote in the process of sexual reproduction).
GAMETE: reproductive cell capable of fusing in pairs to form new individual (zygote).
GAMONT: generation which forms gametes in sexual reproduction, commonly with megalospheric test (A-forms).
GERM DENTICLE: these are suppressed or aborted denticles, which could not develop into mature structures owing to crowded condition along growing edge of conodont.
GIRDLE (Diatoms): also called the connecting band or the girdle band. It holds the epivalve and hypovalve together. (Dinoflagellates: a term which can be synonymous with the transverse furrow or cingulum of a motile dinoflagellate or the paracingulum of a dinocyst. Its use is not recommended.)
GOLGI APPARATUS: local structure in cytoplasm of cells. Appears like a group of membranes constituting a number of flattened sacs aligned roughly parallel. Secretory function of wall material.
GONAL: the point of intersection of three or more paraplates, or of paraplates and paracingulum or parasulcus.
GONYAULACACEAN: dinoflagellates with the *Gonyaulax* tabulation, or dinocysts with the *Gonyaulacysta* paratabulation.
GRANULAR HYALINE WALL: perforate, lamellar test composed of granular calcite, seen between crossed nicols as multitude of tiny flecks of color.
GROWTH LAMELLAE: numerous thin layers or sheaths forming a conodont, and its basal body. They are accreted around a growth center.
GYMNODINIOID STAGE: the thin-walled uniflagellate stage which is formed soon after release of the protoplast from the dinocyst on excystment, and which rapidly gives away to a biflagellate stage.
GYROGONITES: the spiral-shaped calcified portion of the oogonium (charophyta).

HAPLOID STAGE: having a single set of unpaired chromosomes in each nucleus (e.g. in gametes).
HAPTONEMA: flagellar apparatus in coccolithophores.
HEIGHT: distance between two planes perpendicular to shell axis.
HEMISOLENIC: type of carapace closure having at each valve both salient (positive) and depressed (negative) elements.
HETEROCOCCOLITHS: coccoliths composed of crystallites of different shapes and sizes.
HETEROMORPH: term used in dimorphic species of extinct ostracodes for adult carapaces of that sex which differ more from the larval carapaces than those of the other sex. Considered to represent females.
HETEROMORPHIC: processes of more than one type in a single dinocyst or acritarch.
HETEROPHASIC: containing two or more distinct phases, e.g., alternation of generations produces heterophasic life cycle.
HETEROTROPHIC: organisms requiring organic food supply form its environment.
HINGE: dorsal part of the valves serving for their articulation.
HINGE LINE: line along which the valves articulate, seen in closed carapace.
HISTIUM: a frill-like adventral structure occurring dorsally from the velum.
HOLOCOCCOLITHS: coccoliths composed of crystals of same shape and size.
HOLOPLANKTON: organisms living their complete life cycle in the floating state.
HOLOSOLENIC: type of carapace closure having all the elements of both cardinal and extracardinal margins in the same valve either depressed or raised.
HOMOMORPHIC: when the processes on a dinocyst or acritarch are all of the one type, i.e. are all the same.
HORN: an outbulge or extension of the wall, or of its outer layer, in motile dinoflagellates or dinocysts. Horns may be apical, lateral or antapical in position. (In nassellarian radiolarians, a term sometimes applied to a stout apical spine.)
HYALINE AREA: usually refers to an area on the frustule which is composed of clear silica with no structures.
HYPOCYST: the portion of a dinocyst corresponding to the hypotheca of a motile dinoflagellate. In paratabulate dinocysts this is the area posterior to the paracingulum and consisting of the postcingular, posterior intercalary (if present) and antapical paraplates. Most or all of the parasulcus is located on the hypocyst.
HYPOCYSTAL ARCHEOPYLE: an archeopyle formed from the loss of some of the paraplates of the hypocyst.
HYPOSOME: in naked or unarmored motile dinoflagellates, the portion of the amphiesma which is posterior to the cingulum.
HYPOTHALLIUM: the basal or inner tissue in a coralline alga.
HYPOTHECA: also called the hypovalve. The smaller of the two valves in the frustule of a diatom. In a thecate motile dinoflagellate, the portion of the theca which is posterior to the cingulum. It includes the postcingular, posterior intercalary (if present) and antapical plates. Most or all of the sulcus is located on the hypotheca.

IMPERFORATE: without pores, used for porcelaneous tests and in describing ornamentation.
INAPERTURATE: without clearly defined apertures.
INFRALAMINAL ACCESSORY APERTURE: opening in planktonic foraminiferal test leading to cavity beneath accessory structures (bullae) at the margin of these structures.
INNER LAMELLA: thin layer enveloping the ostracode body in anterior, ventral and posterior parts of the carapace; chitinous, with marginal parts often calcified.
INTECTATE: without a tectum, that is, no ektexinous sculpture elements fused to form a membrane, or tectum, outside the endexine.
INTERCALARY: located either between the apical and the precingular or between the postcingular and antapical series of plates in the theca, or between the corresponding series of paraplates in the dinocyst.
INTERCALARY ARCHEOPYLE: an archeopyle formed from the loss of one, or more, of the anterior intercalary paraplates.
INTERCALARY BAND: the area between adjacent plates in peridinioid dinoflagellate thecae. Also found in cysts. It is commonly striate.

INTERGONAL: synonymous with parasutural.

INTRALAMINAL ACCESSORY APERTURE: opening in planktonic foraminiferal test leading through accessory structures (bullae) into cavity beneath but not directly into chamber cavity.

INTRAUMBILICAL APERTURE: opening of test located in umbilicus but not extending outside of it.

INTRATABULAR: within the parasutures of individual paraplates in dinocysts.

INVOLUTE: strongly overlapping; in enrolled forms, later whorls completely enclosing earlier ones.

ISOTONIC: solutions having the same osmotic pressure.

ITERATIVE EVOLUTION: the repeated evolution of a morphologic feature in related taxa.

KALYPTRA: the outer flocculant, granular, organic mantle found in some dinocysts such as the genus *Kalyptea*.

KENOZOOID: zooid in bryozoan colony modified for supporting functions.

LABYRINTHIC: having complex spongy wall with interlaced dendritic channels perpendicular to surface, characteristic of some agglutinated foraminifers.

LAMELLAR: composed of thin layers of aragonite or calcite, one layer being formed with addition of each new chamber and covering whole previously formed test.

LANCEOLATE: flat, narrow, and tapering to point.

LAST WHORL: in coiled shells, last-formed complete volution.

LATERAL PORE CANAL: tubular passage traversing more or less perpendicularly the outer lamella. Syn.: normal pore canal.

LATTICE, LATTICED, LATTICEWORK: referring to a type of wall structure in radiolarians consisting of a regular open two-dimensional network of connecting bars.

LENTICULINE: lens-shaped.

LIGAMENT: narrow stripe of cuticle uniting the two valves along the dorsal margin.

LIMBATE: referring to thickened border of a suture, may also be elevated.

LINEAR PROCESS COMPLEX: a series of processes arranged in a straight line, linked or unlinked proximally, along their length, or distally.

LINE OF CONCRESCENCE: proximal line of junction of the outer and inner lamellae.

LIP: elevated border of aperture, may be at one side of aperture or completely surround it.

LIST: low ridge along the valve margin of some podocopids, situated proximally to the selvage.

LOBE: major elevation usually best developed in the dorsal part of the ostracode valve and reflected on its internal surface. (In conodonts: lateral extension of platform or blade of a conodont.)

LOCULAR WALL: cell wall composed of two layers of silica separated by vertical sections, also made of silica.

LONGITUDINAL FLAGELLUM: the flagellum which arises from the posterior sulcal pore and is directed backwards along the longitudinal furrow or sulcus in a motile dinoflagellate.

LONGITUDINAL FURROW: the furrow on the ventral surface of the motile dinoflagellate which houses the longitudinal flagellum. It lies wholy or partially on the hyposome or hypotheca. Synonymous with sulcus.

LOPHOPHORE: circular or horshoe-shaped ring around mouth of bryozoans bearing ciliated tentacles.

LUNARIUM: horse-shoe shaped projection on proximal side of aperture in cystoporate bryozoa.

LYSOCLINE: level in the water column where significant amount of carbonate begins to be dissolved.

Ma: mega annum = million years B.P.

MAIN BODY: in chlorate dinocysts the central portion of the cyst from which the processes arise.

MAIN SPINES: conspicuous large, long spines on radiolarian skeletons; usually relatively few in number.

MANDIBULAR SCARS: imprints of chitinous support rods of the mandible, situated in front of the adductor muscle scars and below the frontal muscle scars.

MANTLE: the edge of the valve where it bends to join the connecting girdle.

MARGINAL PORE CANALS: tubules passing through the zone of concrescence. Synonymous with radial pore canals.

MATRIX: a thin inner zone of the radiolarian ectoplasm separating the calymma from the central capsule.

MEDIAN SECTION: slice in central sagittal position, perpendicular to axis of coilings.

MEDIAN SULCUS: sulcus corresponding to the attachment of adductorial muscles, designated as S_2.

MEDULLARY SHELL(S): the inner shell(s) of multi-shelled radiolarians.

MEGALOSPHERIC: having large proloculus, commonly representing gamont generation.

MERODONT: hinge with teeth in one valve only.

MEROPLANKTONIC: near-shore species which spend part of their existence in the pelagic realm and the remainder on the sea floor, probably as a resting spore.

MESOCOEL: the cavity, or cavities, between the endophragm and the mesophragm in a dinocyst.

MESOCYST: the body formed by the mesophragm in a dinocyst.

MESODERM: the middle wall in acritarchs with three walls.

MESOPHRAGM: the middle wall in dinocysts with three walls, in which the outer is the periphragm.

MESOPORE: modified zooid in trepostomatous Bryozoa.

METABOLISM; the process by which nutritive matter is built into living material.

MICROFLAGELLATES: tiny members of the group of flagellated protists.

MICROGRANULAR: microscopically granulose, referring to wall composed of minute calcite crystals, probably originally granular but possibly recrystallized; granules may be aligned in rows perpendicular to outer wall, resulting in fibrous structure.

MICROSPHERIC: having small proloculus, commonly agamont (schizont) generation, adult test large.

MICROSPORE: the small male spore in heterosporous ferns and fern allies, as in *Isoetes, Salvinia, Azolla*, and *Marsilea*; also, in microsporogenesis, the monoculeate forerunner of the pollen grain.

MILIOLINE: formed as in Miliolacea, commonly with elongate chambers, added in differing planes of coiling.

MISCELLANEOUS ARCHEOPYLES: archeopyles that cannot be placed into one of the four major categories which are the apical, intercalary, precingular and combination groups

MITOCHONDRIA: rod or thread-like microscopic bodies occurring in cytoplasm of every cell (except bacteria and blue-green algae).

MONAD: a pollen grain occurring singly, as opposed to pollen in tetrads or polyads.

MONOCOLPATE: with a solitary colpus.

MONOCOLPORATE: with one colpus interrupted by a pore.

MONOLETE: with a linear tetrad scar.

MONOPORATE: with one pore.

MOTILE DINOFLAGELLATE: a dinoflagellate that has not encysted, but is a functional component of the plankton, swimming, feeding and reproducing.

MULTILOCULAR: many-chambered test.

MUSCLE SCAR: marking on internal surface of the valve which indicates place of muscle attachment; it is depressed or raised and shows a different shell structure than the rest of the valve.

NONVASCULAR: referring to plants without conducting tissues or well-differentiated roots, stems, and leaves.
NORMAL PORE CANAL: see lateral pore canal.
NUCLEOLI: small dense bodies which contain ribose nucleo-protein.
NUCLEUS: spherical, compact mass of chromatin surrounded by membrane, lying within cytoplasmic body.

OBLATE: shape when the P:E ratio is 0.50:0.75.
OBLATE SPHEROIDAL: shape with a P:E ratio of 0.88:1.00.
OBLIQUE SECTION: slice through test cut in direction neither parallel to axis of coiling nor normal to it.
OLIGOTYPIC: containing one type only - assemblage of only one species.
OOGONIUM: the female reproductive organ of the Charophyta.
OOZES: ocean bottom sediments (usually deep) consisting mainly of skeletal material, e.g. diatom ooze, ratiolarian ooze, foram ooze, nannoplankton ooze.
OPERCULUM: the paraplate or group of paraplates, which are lost or partially detached in archeopyle formation and which are bounded by the principal archeopyle suture. Chitinous lid covering the aperture in cheilostomatous Bryozoa. Corneous or calcareous structure borne by foot of pteropods and serving for closure of aperture, wholly or partly.
ORIFICE: aperture or other opening in test.
OUTER LAMELLA: outer, usually well calcified layer, constituting the whole outer part of the ostracode valve. Along the extracardinal margin of the valve, it passes by a simple bend or by fusion into the inner lamella.

PALMATE: flat, resembling hand with outspread fingers.
PARACINGULAR: adjectival form of paracingulum.
PARACINGULUM: the area on the dinocyst analogous to the cingulum of the motile dinoflagellate. Paraplates may or may not be visible.
PARALLEL SECTION: slice through test in plane normal to axis of coiling but not through proloculus.
PARAPLATE: the cyst equivalent to a plate in the theca.
PARASULCAL: adjectival form of parasulcus.
PARASULCAL NOTCH: the re-entrant angle in the apical archeopyle margin which marks the posterior extension of the first apical paraplate. It is immediately anterior to the parasulcus.
PARASULCAL TONGUE: the posterior extension of the first apical paraplate in the operculum of an apical archeopyle.
PARASULCUS: the area on the dinocyst analogous to the sulcus or longitudinal furrow of the motile dinoflagellate. Paraplates may or may not be visible.
PARATABULATION: the pattern or arrangement of the constituent paraplates in a dinocyst. It is usually expressed as an alpha-numeric formula which gives the series of paraplates present, and the number of paraplates within each series.
P:E RATIO: the relationship between the lengths of the polar and equatorial axes in radiosymmetric grains.
PENNATE: term used in an informal classification of diatoms. Pennate diatoms are those forms which have features or structures running parallel to the transapical axis.
PENETABULAR: linear features which lie immediately interior to the margin of a paraplate.
PERFORATE PLATE: a type of wall structure in radiolarians in which a thin plate-like wall is perforated by pores, as opposed to a latticed wall which is constructed of discrete elements (bars).
PERICOEL: the cavity, or cavities, lying between the endophragm and periphragm, or mesophragm and periphragm, in a dinocyst.
PERICYST: the body formed by the periphragm in a dinocyst.
PERIDINIACEAN: dinoflagellates with the *Peridinium* tabulation, or dinocysts with the *Wetzeliella* paratabulation.
PERIDINIOID: the characteristic outline of a dinoflagellate of the genus *Peridinium*, with a pointed apex or apical horn and with two antapical horns. Also applied to motile dinoflagellates and dinocysts with a pentagonal outline, which may or may not be prolonged into one apical, two lateral (paracingular) and one or two antapical horns.
PERINE: an outer layer in the wall of certain spores, considered by some to be part of the ektexine and by others to be extra-exinous.
PERIPHRAGM: the outer wall in dinocysts with two walls. The outermost wall in dinocysts with three walls, of which the middle is the mesophragm. Next to the outer wall in dinocysts with three walls of which the outermost is the ectophragm.
PERIPORATE: with pores, exceeding three in number, arranged more or less equidistantly on the grain surface.
PERIPROLATE: shape when the P:E ratio is ⟩2.
PERITHALLIUM: the upper or outer layer of tissue in a coralline alga.
PERVALVAL AXIS: also called the cell axis. The axis through the center point of the two valves.
PHAEODIUM: a mass of pigmented spaerules in the ectoplasm of one major group of radiolarians (*Tripylea* or *Phaeodaria*).
PHAGOTROPHY: ingestion of foreign food particles.
PHIALINE: having everted rim on apertural neck, as on neck of a bottle.
PHOTOSYNTHESIS: a process in which carbon dioxide and water are chemically combined to form carbohydrates, the energy for the process being sunlight.
PHRAGMA: the wall of a dinocyst. It can be composed of one, two, or more than two layers.
PIT: a structure around the growth center of a conodont which ceases to develop after it has attained a certain size.
PLANISPIRAL: coiled in single plane.
PLATE: one of the constituent and separable units of the theca.
PLATE FORMULA: the alpha-numeric representation of the plates in a thecate dinoflagellate. It shows the series of plates present, and the number of plates within each series, in an abbreviated formula. Thus, 4′, 1a, 6″, 6c, 6‴, 1⁗ means there are four apical, one anterior intercalary, six precingular, six cingular, six postcingular and one antapical plates in the taxon whose tabulation is given in the plate formula.
POLAR AXIS: imaginary line passing from pole to pole and constituting a major reference of symmetry.
POLLEN: the binucleate or trinucleate male reproductive body in seed plants.
POLYAD: grains united in multiples of more than four.
POLYHEDRON: a solid resulting from the intersection of many plane faces.
POLYMORPHISM: morphologically different forms of same species which may be result of different generations.
POLYPIDE: soft parts of individual in a bryozoan colony.
PORCELANEOUS: having calcareous, white, shiny, and commonly imperforate wall resembling porcelain in surface appearance; shows low-polarization tints between crossed nicols and has majority of crystals with *c*-axes tangential, or more rarely arranged radially; commonly brown in transmitted light.
PORE: a more or less isodiametric germinal aperture in spore or pollen with a length to width ratio of ⟨2. Opening of pore canals in ostracodes.

PORE CANAL: tubular passage traversing the valve wall of ostracodes. The larger of the two types of pores which partially or completely penetrate the wall of a tasmanitid.
PORE (Forminifera): found, slit-like or irregular openings ≅ 5–6 μm in size partially perforating the test wall of foraminifera.
PORTICUS (pl. portici): asymmetrical apertural flaps.
POSTCINGULAR: located between the cingulum and antapex in a motile dinoflagellate, and between the paracingulum and antapex in a dinocyst.
POSTCINGULAR HORN: a horn arising from a position on the hypotheca of a motile dinoflagellate immediately posterior to the cingulum, i.e. from a postcingular plate. Alternatively a horn arising from a position on the hypocyst of a dinocyst immediately posterior to the paracingulum, i.e. from a postcingular paraplate.
POSTCINGULAR PARAPLATE: one of the latitudinal series of paraplates lying immediately posterior to the paracingulum in a paratabulate dinocyst.
POSTCINGULAR PLATE: one of the latitudinal series of plates lying immediately posterior to the cingulum in a tabulate motile dinoflagellate.
POSTEQUATORIAL: in acritarchs one of the series of paraplates lying immediately posterior to the equator or "paracingulum". In dinocysts synonymous with postcingular.
POSTERIOR INTERCALARY PARAPLATE: the paraplate, or one of the group of paraplates, of a paratabulate dinocyst, which lies between the postcingulars and antapical(s), without touching the cingulum or antapex.
POSTERIOR SULCAL PORE: the pore from which the longitudinal flagellum arises.
PRECINGULAR: located between the apex and cingulum in a motile dinoflagellate, and between the apex and paracingulum in a dinocyst.
PRECINGULAR ARCHEOPYLE: an archeopyle formed from the loss of one, or more, of the precingular paraplates.
PRECINGULAR PARAPLATE: one of the latitudinal series of paraplates lying immediately anterior to the paracingulum in a paratabulate dinocyst.
PRECINGULAR PLATE: one of the latitudinal series of plates lying immediately anterior to the cingulum in a tabulate motile dinoflagellate.
PRE-EQUATORIAL: in acritarchs one of the series of paraplates lying immediately anterior to the equator or "paracingulum". In dinocysts synonymous with precingular.
PRIMARY APERTURE: main opening of test.
PRINCIPAL ARCHEOPYLE SUTURE: the suture developed between the operculum and archeopyle margin, or within the operculum when it results in complete separation of the portions.
PROCARYOTES: morphologically primitive organisms (bacteria and blue-green algae; mostly single cells or simple filaments) which do not have DNA separated from the cytoplasm by an envelope.
PROCESS: an essentially columnar or spine-like projection arising from the surface of a dinocyst or acritarch. Processes may be simple or intricately branched and interconnected. Processes are rarely, if ever, found in motile dinoflagellates.
PROCESS COMPLEX: the association of three or more adjacent intratabular processes to form a distinctly arranged and aligned group, often united, proximally, along their length and/or distally.
PROCESS FORMULA: the alpha-numeric representation of the processes in a dinocyst or acritarch. It shows the series of paraplates present, and the number of paraplates within each series, but only if such paraplates bear processes. Thus 4′, 6″, 6‴, 1p, 1″″ means there are four apical, six precingular, six postcingular, one posterior intercalary and one antapical processes in the taxon represented by this process formula. Since the paracingulars are devoid of processes they are not included.
PROLATE: shape with a P:E ratio of 1.32:2.
PROLOCUS (pl. proloculi). Initial chamber of foraminiferal test.
PROTOPLAST: the protoplasm of the cell excluding the cell wall.
PROXIMAL: side facing inward.
PROXIMATE: a dinocyst which in form and size closely resembles the corresponding motile stage, which develops from it upon excystment.
PSEUDONODULE: also called a pseudo-ocellus. A small clear area, usually on the periphery of the valve, slightly raised and frequently surrounded by small pores.
PSEUDOPODIA: cytoplasmic projections serving for locomotion, attachment, and capture of food in foraminifers.
PSEUDOPYLOME: a false pylome found in the antapical region of some acritarchs.
PSEUDORAPHE: a narrow, hyaline area running parallel to the apical axis. It is not true raphe because there is no cleft or notch on the valve.
PUNCTA (pl. punctae): holes or perforations.
PYLOME: the circular excystment opening found in many acritarchs.

RADIAL BEAM: a radial connecting bar in radiolarian skeletons.
RADIAL CANAL: the smaller of the two types of pores which partially or completely penetrate the wall of a tasmanitid.
RADIAL MICROSTRUCTURE: calcareous tests consisting of calcite or aragonite crystals with *c*-axes perpendicular to surface; under crossed nicols shows black cross with concentric rings of color mimicking negative uniaxial interference figure.
RADIAL PORE CANAL: see marginal pore canal.
RADIATE APERTURE: opening associated with numerous diverging slits.
RADIOLARIAN EARTH: a soft, fine-grained, porous rock similar to diatomite but with radiolarians dominant.
RADIOLARITE: a hard siliceous rock rich in radiolarian remains, term generally restricted to cherts or similar dense lithologies.
RADULA: ribbon-like band containing teeth in the pharynx of pteropods.
RAMUSCULOSUM TYPE: synonym of branched process.
RAPHE: a line or slit found in some pennate diatoms which runs parallel to the apical axis. In side view the raphe may be V-shaped or notched.
RECTILINEAR: growing in a straight line.
REFLECTED TABULATION: synonym of paratabulation.
RELICT APERTURES: short radial slits around umbilicus of test which remains open when umbilical portions of equatorial aperture are not covered by succeeding chambers.
RESTING CYST: synonym of dinocyst in dinoflagellate usage.
RETICULATE: ornamental ridges at surface of test or inner meshwork.
RHIZOPODIA: bifurcating and anastomosing pseudopodia.
ROSTRUM: a beak-like projection at the anterior carapace end of many myodocopids, overhanging a gap between the valves, called rostral incisure.
RUGOSE SURFACE: rough irregular ornamentation, may form ridges.

SACCATE: possessing sacci.
SACCUS (pl. sacci): air sac, bladder, or wing characteristic of vesiculate pollen types in which the ektexine and endexine have become separated; also hollow projections of the walls of some fern spores.
SAGITTAL RING: in nassellarians, a skeletal structure consisting of median bar and apical and vertical spines connected by an arched bar.

SAGITTAL SECTION: slice through test perpendicular to axis of coiling and passing through proloculus.

SCULPTURAL ELEMENT: a projection of less than 5 μm height on the outer surface of an acritarch.

SCULPTURE: ornamentation, or the external appearance of the ektexine without reference to its makeup or construction.

SECONDARY APERTURES: additional or supplementary openings into main chamber cavity; areal, sutural, or peripheral in position.

SECONDARY PIT CONNECTION: narrow passage connecting adjacent cells not belonging to the same filament in tissue of coralline red algae.

SELVAGE: chitinous fringe with calcified base, developed along the extracardinal margin and serving to seal the closed valves.

SEPTULUM (pl. septula): ridge extending downward from lower surface of spirotheca so as to divide chambers partially.

SEPTUM (pl. septa): partition between chambers, commonly consisting of previous outer wall or apertural face, may have single layer, be secondarily doubled enclosing canal systems or be primarily double. A membraneous, linear projection on the wall of a dinocyst or acritarch. Commonly parasutural in position.

SESSILE: attached, sedentary.

SEXUAL DIMORPHISM: difference in morphology of sexes of the same species.

SIEVE PLATE: a thin siliceous plate covering the areolae and perforated by fine pores.

SIEVE PLATE: minute discoidal plate with numerous circular, triangular, and polygonal micropores arranged in concentric rows, contained in pore canal of certain foraminifers.

SIMPLE HETEROMORPHIC: when there are branched processes only in a single cyst.

SIMULATE PROCESS COMPLEX: an arrangement of intratabular processes in the form of a closed polygon, linked or unlinked proximally, along their length, or distally. The complex is developed within, but parallel to the boundaries or parasutures of a paraplate. Partly synonymous with penetabular.

SINISTRAL: shell arranged as in mirror image of dextral (see dextral).

SIPHON: internal tube extending inward from aperture.

SOLEATE PROCESS COMPLEX: a series of intratabular processes arranged in a horse-shoe or crescent on an individual paraplate. The processes of an individual complex may be united proximally, along their length, or distally.

SPHEROIDAL: shape like a sphere with a P:E ratio of 0.88:1.14.

SPICULE: referring to a radiolarian skeleton or skeletal segment consisting of a simple association of radiating spines.

SPINOSE: having fine elongate solid spines on surface of test.

SPIRAL SIDE: part of test where all whorls are visible, also called dorsal side.

SPIRILLINE: planispiral nonseptate tube enrolled about globular proloculus.

SPIROTHECA: outer or upper wall of test in fusulinaceans.

SPIROUMBILICAL: interiomarginal aperture extending from umbilicus to periphery and onto spiral side.

SPONGY: referring to a type of wall structure in radiolarians consisting of an irregular three-dimensional system of interconnecting bars.

SPORANGIUM: a cell which eventually produces one or more spores.

SPORE: as used here, the reproductive body in the Pteridophyta that is asexual or sexual, isosporous plants producing asexual isospores and heterosporous plants forming male microspores and female megaspores.

STOLON: tube-like projections connecting chambers in orbitoids.

STREAMING: continuous linear motion in gel-like protoplasm, particularly pseudopodia.

STREPTOSPIRAL: coiled in several planes.

STRIATE: marked by parallel grooves or lines.

SUBOBLATE: shape with a P:E ratio of 0.75:0.88.

SUBPROLATE: shape when the P:E ratio is 1.14:1.33.

SULCAL PARAPLATE: a paraplate located within the parasulcus of a dinocyst.

SULCAL PLATE: a plate located within the sulcus of a motile dinoflagellate.

SULCAL PORE: one of the two pores located in the sulcus and from which arise either the transverse flagellum or the longitudinal flagellum.

SULCUS: synonym of longitudinal furrow in dinoflagellates. A trench or depression on the valve surface, roughly dorsoventrally oriented in ostracodes.

SUPPLEMENTARY APERTURES: secondary openings in test additional to primary aperture or completely replacing primary aperture.

SUPPLEMENTARY MULTIPLE AREAL APERTURES: subordinate openings in tests.

SUTURAL SUPPLEMENTARY APERTURES: relatively small sutural openings which may be single or multiple, with many openings along the sutures.

SUTURE: line of union between two chambers or between two whorls.

SWARMERS: flagellated gamete cells produced by protozoans.

SYMBIOSIS: life association mutually beneficial to both organisms; commonly refers to green or blue-green algae or yellow cryptomonads.

TABULATE: composed of plates.

TABULATION: the pattern or arrangement of the constituent plates in a tabulate motile dinoflagellate. It is usually expressed as an alpha-numeric formula which gives the series of plates present, and the number of plates within each series.

TANGENTIAL SECTION: slice through test parallel to axis of coiling or growth but not through proloculus.

TECNOMORPH: term used in dimorphic species of extinct ostracodes for larval carapaces and those adult carapaces which are essentially similar to the larval ones, generally interpreted as belonging to adult males.

TECTIN: organic substance having appearance of chitin but distinct chemically.

TECTUM: an outer layer of the wall formed by the distal fusion of ektexinous sculpture elements; grains with a more or less complete tectum are tectate, those with a partial covering of tectum are classified as semitectate.

TEST: shell or skeletal covering, may be secreted, gelatinous, chitinous, calcareous or siliceous; or composed of agglutinated foreign particles, or combination of these. In acritarchs all of the cyst, i.e. vesicle plus processes.

TETHYS: ancient circum-global equatorial seaway.

TETRAD: pollen or spore type in which the components are four in number; tetrads occur in the form of a tetrahedron (tetrahedral), rhomboid (rhomboidal).

TETRATABULAR: when there are four adjoining paraplates in an operculum, which is usually from an apical archeopyle.

THALLUS: plant body without true roots, stems, and leaves.

THANATOCOENOSES: death-assemblage accumulation of dead organisms which did not necessarily live together.

THECA: the body formed by the plates of a thecate or tabulate motile dinoflagellate.

THECATE: composed of plates which together form a theca; or possessing a theca.

THORAX: in nassellarian radiolarians, the second (first postcephalic) chamber or segment.

TISSUE: a group of cells of similar structure which performs a specialized function.
TOOTH: projection in aperture of test, may be simple or complex, single or multiple.
TOOTH PLATE: internal, apertural modification consisting of plate that extends from aperture through chamber to previous septal foramen. One side may be attached to chamber wall or base attached to proximal border of foramen.
TRANSAPICAL AXIS: the transverse axis of the valve.
TRANSAPICAL SUTURE: the excystment suture developed in peridinioid dinocysts with a transapical excystment aperture. It follows the ventral margins of paraplates 3″, 1a, 3′, 3a and 5″.
TRANSVERSE FLAGELLUM: the flagellum which arises from the anterior sulcal pore and is directed laterally in an equatorial position along the transverse furrow or cingulum, in a motile dinoflagellate.
TRANSVERSE FURROW: the equatorially aligned furrow which almost completely encircles the amphiesma and which houses the transverse flagellum. It separates the episome or epitheca from the hyposome or hypotheca. Synonymous with cingulum.
TRICOLPATE: possessing three colpi.
TRICOLPORATE: with three colpi, each containing a pore; when the pores are rather indistinct, the type is referred to as tricolporoidate.
TRILETE: with a triradiate tetrad scar.
TRIMORPHISM: some megalospheric forms were plurinucleate and reproduced, producing a third generation.
TRIPOLI: a commercial term for soft, fine-grained porous material similar to or identical with diatomite or radiolarian earth.
TRIPORATE: with three pores.
TRISERIAL: chambers arranged in three columns, three chambers in each whorl.
TRISPINOSUM TYPE: a process that is simple, or bifurcate.
TROCHOID: Trochospiral, rotaloid, rotaliform; chambers coiled spirally, evolute on one side, involute on other.
TROCHOSPIRAL: trochoid, rotaliform; spirally coiled chambers, evolute on one side of test, involute on opposite side.
TUNNEL: resorbed area at base of septa in central part of test in many fusulinids, facilitating communication between adjacent chambers.
TYCHOPELAGIC: species which spend most or all of their lives on near-shore sea bottoms.

UMBILICAL SIDE: involute side in trochospiral forms, with only chambers of final whorl visible around umbilicus.
UMBILICAL TEETH: triangular modification of apertural lip, of successive chambers in forms with umbilical aperture giving characteristic serrate border to umbilicus.
UMBILICATE: having one or more umbilici.
UMBILICUS (pl. umbilici): space formed between inner margins of umbilical walls of chambers belonging to same whorl. Pteropods: cavity or depression formed around shell axis between faces of adaxial walls of whorls where these do not coalesce to form a solid columella; in conipiral shells its opening is at the base of shell.
UNARMORED: naked.
UNICELL: an organism consisting of a single cell.
UNIFLAGELLAGE STAGE: the protoplast which is surrounded by a delicate membrane and is released from the dinocyst on excystment. It is uniflagellate and lacks both a transverse furrow and a longitudinal furrow. It is the initial free-swimming stage which lasts approximately 15 min before giving way to the gymnodinioid stage.
UNILOCULAR: single-chambered.
UNIPARTITE HINGE: hinge not subdivided into median and terminal elements.
UNISERIAL: having chambers arranged in a single row.

VACUOLE: globular inclusion in cytoplasm; includes contractile vacuoles, food vacuoles.
VALVE: the largest component of a frustule. The two valves of a frustule fit over one another much like a pill box.
VALVE CAVITY: space between the outer and the inner lamellae of the valve.
VEGETATIVE DIVISION: asexual reproduction.
VEGETATIVE STAGE: all stages in the life cycle except the encysted stage or dinocyst. Partly synonymous with motile dinoflagellate.
VELUM: adventral elongate frill-like structure which parallels the extracardinal margins in some beyrichiomorphs.
VENTRAL: lower side of test, commonly used for umbilical side; opposite to dorsal; commonly apertual side. In dinoflagellates: the surface containing, the sulcus in the motile dinoflagellate, and the parasulcus in the dinocyst.
VENTRAL NOTCH: synonymous with parasulcal notch.
VENTRAL PARAPLATE: synonymous with sulcal paraplate.
VENTRAL TONGUE: synonymous with parasulcal tongue.
VESICLE: the test of an acritarch excluding the processes. Frequently called central body.
VESICULATE: possessing sacci.
VESTIBULE: space between the outer lamella and the calcified part.
VESTIBULUM (pl. vestibula): compartment situated between the exopore and endopore and resulting from the differentiation of the ektexine and endexine in the pore area.
VIBRACULARIA: zooid with the operculum modified into a long seta in cheilostomatous Bryozoa.
VIRGATODONT: terms used for hinges of Paleozoic ostracodes in which longitudinal grooves and bars prevail.
VISBYENSE TYPE: in acritarchs broad conical to subconical processes which distally are minutely branched.
VISCERAL MASS: in pteropods, mass in which internal organs are concentrated.

WATER MASS: a parcel of water of suboceanic scale with a distinctive set of physical-chemical characteristics usually defined on the basis of temperature and salinity.
WHORL: single turn or volution of coiled test (through 360°).
WINGS: in nassellarian radiolarians a term sometimes applied to conspicuous pre-terminal spines directed away from the apex, usually homologs of dorsal and primary lateral spines.

XANTHOSOME: small brown or yellowish, globular inclusions in cytoplasm.

ZOARIUM: term for the bryozoan colony.
ZONE OF CONCRESCENCE: zone in which the outer and inner lamellae fuse by their lateral surfaces. Syn.: zone of fusion.
ZOOECIUM: skeleton of individual zooid in bryozoan colony.
ZOOID: the single individual in a bryozoan colony.
ZOOSPORANGIUM: a body within which are produced asexual spores termed zoospores.
ZOOSPORE: a naked asexual spore possessing one or more flagella.
ZOOXANTHELLAE: symbiotic cells (usually yellow pigmented) of algal origin.
ZYGOTE: result of fusion of two gametes in process of sexual reproduction.
ZYGOTIC CYST: a cyst which contains a fertilized ovum.

INDEX